Recent Trends in Sustainability and Management Strategy

Conference Proceedings of
Fourth International Conference on

Recent Trends in Sustainability and Management Strategy

ICSMS–2015

EDITORS

Dr. V.S. Gajavelli
Dr. Kapil Chaturvedi
Dr. Abhishek Narain Singh

September 2015
IMT Nagpur

**Institute of
Management Technology**
Nagpur

● ● ● ●

Conference Partners

CALIFORNIA STATE UNIVERSITY
SAN BERNARDINO

NEERI

ALLIED PUBLISHERS PVT. LTD.

New Delhi • Mumbai • Kolkata • Lucknow • Chennai
Nagpur • Bangalore • Hyderabad • Ahmedabad

ALLIED PUBLISHERS PRIVATE LIMITED

1/13-14 Asaf Ali Road, **New Delhi**–110002
Ph.: 011-23239001 • E-mail: delhi.books@alliedpublishers.com

87/4, Chander Nagar, Alambagh, **Lucknow**–226005
Ph.: 0522-4012850 • E-mail: appltdlko9@gmail.com

17 Chittaranjan Avenue, **Kolkata**–700072
Ph.: 033-22129618 • E-mail: cal.books@alliedpublishers.com

15 J.N. Heredia Marg, Ballard Estate, **Mumbai**–400001
Ph.: 022-42126969 • E-mail: mumbai.books@alliedpublishers.com

60 Shiv Sunder Apartments (Ground Floor), Central Bazar Road,
Bajaj Nagar, **Nagpur**–440010
Ph.: 0712-2234210 • E-mail: ngp.books@alliedpublishers.com

F-1 Sun House (First Floor), C.G. Road, Navrangpura,
Ellisbridge P.O., **Ahmedabad**–380006
Ph.: 079-26465916 • E-mail: ahmbd.books@alliedpublishers.com

751 Anna Salai, **Chennai**–600002
Ph.: 044-28523938 • E-mail: chennai.books@alliedpublishers.com

Hebbar Sreevaishnava Sabha, Sudarshan Complex-2
No. 24/1, 2nd Floor, Seshadri Road, **Bangalore**–560009
Ph.: 080-22262081 • E-Mail: bngl.books@alliedpublishers.com

3-2-844/6 & 7 Kachiguda Station Road, **Hyderabad**–500027
Ph.: 040-24619079 • E-mail: hyd.books@alliedpublishers.com

Website: www.alliedpublishers.com

ISBN: 978-93-85926-55-6

Published by Sunil Sachdev and printed by Ravi Sachdev at Allied Publishers Pvt. Ltd., (Printing Division), A-104 Mayapuri Phase II, New Delhi-110064

Foreword

Dear ICSMS–2015 Conference Delegates,

On behalf of Institute of Management Technology Nagpur, I take immense pleasure in welcoming you all to our *4ᵗʰ International Conference on Sustainability and Management Strategy*, to be held in Nagpur. This year the conference theme will be sustainability and management strategy in various related areas of business management and other disciplines. Further, authors are encouraged to contribute on two special tracks: Sustainable Business Models and Sustainability Reporting. Interestingly, majority of the research contributions are on the main theme of the conference and that shows the significance of this growing area for all the stakeholders of the society.

With enterprises world over redefining the very purpose of their existence from profit-making to include societal and environmental responsibilities, sustainability has taken a centre stage for these business entities. Governments, not-for-profit organizations and social activist groups are also equally concerned about the overall impact of economic development on environmental and societal fabric. It is in this context, that ICSMS–2015 aims at sharing research and experience based knowledge among researchers, academicians, policy-makers, industry veterans, NGOs, consultants, students and practitioners in their respective fields.

The insights from distinguished keynote speakers, industry consultants and research scholars are of immense value for society in general and business, academia and management students in particular. I earnestly hope that the two-day conference deliberations and inputs from experts generate more potential ideas for future consideration and bridge the gap between the sustainability as an ever-growing phenomenon and the expected actions on the part of the business organizations and governments at the policy level. Such a fusion is bound to facilitate shaping policies and practices for the growth and development of sustainable societies. Welcome to all the contributors, academic leaders, industry professionals and student researchers.

At the Institute of Management Technology (IMT) Nagpur, all of us feel honored in welcoming you to this conference. The very fact that most of you have travelled from across the country to be here at this conference validates the vitality of the theme.

I am very much delighted to participate in the deliberations and have strong conviction that some of the ideas that may emerge during the two-day deliberations will be pursued further and play an instrumental role in providing solutions to some of the current issues and lead to higher order of enquiry and knowledge creation.

Dr. Subhajit Bhattacharyya
Director Institute of Management Technology, Nagpur

Preface

The International Conference on Sustainability and Management Strategy (ICSMS 2015) was organized at the Institute of Management Technology, Nagpur.

Proceedings of the International Conference on Sustainability and Management Strategy (ICSMS–2015) contains the full text of the papers presented at the conference on 4th & 5th Sept. 2015.

I sincerely thank the conference partner Institutions California State University San Bernardino (CSUSB), USA and Council of Scientific and Industrial Research-National Environmental Engineering Research Institute (CSIR-NEERI), Nagpur.

On behalf of the Organizing Committee, I would like to express my deep appreciation for the interest and involvement extended by all ICSMS–2015 participants. I feel very grateful to you for taking time out of your busy schedule to contribute and enrich the conference deliberations. Without your support and cooperation, I doubt we would have been able to organize the conference in such a proficient and grandeur manner.

I thank the distinguished speakers Mr. Vijay Pratap Singh Aditya, CEO-Ekgaon-One Village One World Network, New Delhi; Dr. Breena Coats, Professor of Global Strategy and Sustainability & Chairman-Management Studies, California State University San Bernardino (CSUSB), USA, Prof. P.S. Dutt, Chief Scientist, Council of Scientific and Industrial Research-National Environmental Engineering Research Institute (CSIR-NEERI), Nagpur; Mr. Dinesh Pillai, CEO, Mahindra Special Services Group (SSG), Mumbai; Dr. C.V. Chalapati Rao, formerly Chief Scientist (CSIR–NEERI), Nagpur for their valuable contributions.

I extend my sincere gratitude to Dr. Subhajit Bhattacharyya, Director, IMT-Nagpur; S.R. Wate, Director, CSIR–NEERI, Nagpur; Dr. Vipin Gupta, Associate Dean and Dr. Breena Coates, Chairperson-Management Studies, CSUSB, USA for their sustained encouragement and support.

I also thank the members of the academic advisory, organizing committee, reviewers, session chairs, authors, fellow colleagues and administrative staff of IMT and the partner Institutions for making the conference a grand success.

Dr. V.S. Gajavelli
Chairman, ICSMS-2015 Professor and Chairperson,
Centre for Sustainability Growth and Development
Institute of Management Technology, Nagpur

Contents

Sustainability of Environmentally Sound Technologies Using Interpretive Structural Modeling

Abhishek Behl[1] and Abhinav Pal[2]

[1]Symbiosis School of Economics, Symbiosis Center for Waste Management and Sustainability, Symbiosis International University
[2]Symbiosis School of Economics, Symbiosis International University
E-mail: [1]abhishekbehl27@gmail.com; [2]abhinav.pal@sse.ac.in

ABSTRACT: *The rapidly growing Indian economy has transformed into a sustainable arena where every player needs to contribute to maintain the equilibrium. Sustainable practices by organizations have contributed towards development. There is a limited contribution from individuals towards the same. This study aims at exploring the key barriers which lead to a limited contribution from individuals residing in rural areas. The limited study in the area of Environmentally Sound Technology (EST) in rural India makes it important as the results could be generalized for a larger community. The barriers found out are diverse ranging from awareness of EST to humanitarian supply chain. The study uses Interpretive Structural Modeling technique to understand the barriers from the individual's perspective. The study projects to build a conceptual framework with linkages between the barriers. The linkages helped in understanding the flow of association between the barriers which would help in dealing with them in a structured approach. The results signify that awareness and perception of individuals towards these technologies are the prime reasons of discontinuity in contributing towards a sustainable supply chain. It was also found out that tradeoff between conventional technologies and EST was not balanced. Public–private partnership in launching these technologies in rural areas is a missing link found out from the research. The results give a fairly clear picture in understanding the barriers and their hierarchy. Therefore, the results will interest both private and public sector in enhancing the diffusion of these green technologies in rural areas. The results give clear directions to each of the stakeholders to understand their role wherever required in bridging the missing links in the sustainable supply chain. The study is related to Indian context which restricts generalizing them for every country. A similar study can be performed for other country's perspective as well.*

Keywords: Sustainability, Environment Sound Technology, Interpretive Structural Modeling, Rural, India.

1. INTRODUCTION

Green sustainable technology is one of the emerging areas covering the periphery of sustainable development and green technology. The amalgamated terminology has evolved from the contribution of research in the fields of sustainable development and renewable sources of energy. In the light of sustainable development in various aspects, the role of technology seems important. It is also worthwhile to understand the level of diffusion of such technologies to understand the worth of the each element that contribute towards such development. The sensitivity of environment with respect to human intervention has slowly started depleted the natural resources and the impact is alarming. The theory of green technology is not new, yet its practical applications are not diverse. The "green technology" is not a concept that has yet to enjoy widespread agreement among economists or environmentalists or an international consensus. It is an extremely complex concept and it is unlikely there can be a consensus on its meaning, use and usefulness and policy implications, in the short term. A "green technology" gives the impression of a technology that is environmentally-friendly, sensitive to the need to conserve natural resources, minimizes pollution and emissions that damage the environment in the production process, and produces products and services the existence and consumption of which do not harm the environment.

It is against such a background that the use of environmentally sound technologies was recognized by the United Nations Conference on Environment and Development (UNCED) as crucial in achieving sustainable development. Chapter 34 of Agenda 21 which deals with environmentally sound technology stresses the 'need for favorable access to and transfer of environmentally sound technologies, in particular to developing countries, through supportive measures that promote technology co-operation and that should enable transfer of necessary technological know-how as well as building up of economic, technical, and managerial capabilities for the efficient use and further development of transferred technology'.

The consumption of perishable natural resources in a highly sustainable manner and release of toxic emissions has led to a great degradation of the environment. The pollution and degradation has also affected environment, economy and society as a whole (Sangwan, 2011). The rising world population and the improving living standards in developing countries have put pressure on the technology industry to grow and transform into being efficient and sustainable. There is a strong need, particularly, in emerging and developing economies to improve technological performance so that there is less pollution, less material and energy consumption, less wastage, etc. One such potential system is environmentally sound technology (EST). It consists of methods and tools to achieve sustainable technology through process optimizations with environmental costs in mind (IEA, 2007). This paradigm shift to newer technology alternatives is urgently required in emerging countries like India to balance their economic growth in tandem with ecological balance. The society is well aware of its responsibility toward environment but there are some factors that hinder the adoption of EST (Singh, 2010).

The theory of environmentally sound technology has evolved over the years. There is a paradigm shift towards protection of environment and contribution of technology to shape the green policies (e.g. Sarkis *et al.,* 2011; Gunasekaran and Gallear, 2012). The output of the researches aims towards creating a equilibrium between humanitarian contribution towards environment both physically and mechanically (Alibeli and Johnson, 2009; Dasgupta *et al.,* 2002; Uzzell and Rathzel, 2008). The topic has been of high importance for researchers in multidisciplinary fields including economics and supply chain. It has also been witnessed that sustainability is a terminology which has been used by every fifth research paper. Irrespective of the kind of diversity prevalent in the research world, there is always scope to investigate fresh perspectives in developing nations. Existing literature has been contributing towards various aspects like green manufacturing (Vachon and Klassen, 2006; Hsu and Hu, 2009; Bai and Sarkis, 2010; Ku *et al.,* 2010; Testa and Iraldo, 2010; Lo *et al.,* 2012), environmental performance (Zhu and Sarkis, 2004; Schoenherr, 2012; Lo *et al.,* 2012), technology transfer (Lema and Lema, 2012; Fu and Zhang, 2011; Munshi, 2004), buying behavior intentions towards green technologies (Young *et al.,* 2010; Rahbar and Abdul, 2011). There are limited studies which relate the behavior intentions of users in rural areas to use environmentally sound technologies. The existing models for EST sparingly build a conceptual framework which would help in analyzing the overall dynamics of the diffusion of technology.

While it is reasonable to believe that environmentally sound technologies will impact growth and contribute towards the economy, it is worthwhile to pause and understand the factors which might slowdown the rate of growth. The present research proposes to study each of the barriers in detail and develop a framework to understand the linkages between them efficiently. Extending the research by Dubey *et al.,* (2014) who proposed the framework thereby relating the leadership, operational practices and developed a sustainable green supply chain, this paper extends in exploring barriers pertaining to the largest stakeholders in the green supply chain *i.e.* users. The barriers may or may not be directly dwelled from users, but it affects them the most (Fischer, 2012; Doss, 2006; Rao and Kishore, 2011). It has been noticed that the researchers have proved the relationships between the dependent and independent variables by collecting data and analyzing results which may be healthy for a small community but largely lack uniformity. With a change in dynamic nature of environment and development of technology, the challenge would always be to understand the rate of diffusion of any technology (Adesina and Zinnah, 1993; Tigabu *et al.,* 2015).

The primary objective is to investigate the barriers which contribute towards the growth and sustainability of environmentally sound technologies in rural India. The identification of barriers is done with a microscopic lens which highlights the challenges faced by the end users. The study would interest stakeholders like technology providers and marketing managers to understand the pulse of diffusion of technology in rural India. Secondary objectives include classifying the variables in the model into categories thereby assigning attributes to them thereby helping researchers and practitioners to understand the correlation between various variables. The variables are derived keeping rural India in mind. India is just a representative developing country in the study and the study could therefore be extended to other parts of the world by making slight alterations in the list of variables.

The research objectives can be therefore structured as follows:

1. To investigate and explore the barriers for diffusion of environmentally sound technology in rural India.
2. To develop a conceptual framework to understand the linkages between the variables using interpretive structural modeling.

2. LITERATURE REVIEW

Environmental Sound Technologies are many in variety and it is therefore important to understand the barriers in adoption of these broadly. This section focuses on identifying relevant literature and performing a structured literature review (Tranfield *et al.,* 2003) to draw barriers related to EST diffusion in rural India. The section briefly explains Environmentally Sound Technology which would standardize the branches and therefore would limit the barriers. There are different aspects, definitions and arguments made by researchers with different school of thought, but the aspects covered in this research broadly cover technology using renewable sources like solar, wind etc.; installed by private or government companies/individuals in rural areas for the benefit of self/society with minimal impact on degradation of the environment (Balan *et al.,* 2007; Deif, 2011; Boltic *et al.,* 2013; Dües *et al.,* 2013).

Studies conducted in the past have focused on issues like technicalities of the technology, marketing and branding of the technology, policy concerns, stakeholder analysis etc. (Luken and Van Rompaey, 2008; UNEP, 2008; Khanna *et al.,* 2002; Zailani *et al.,* 2014). Studies have also indicated the role of private and public sector firms to jointly venture projects to improve efficiency and smoothen the regulatory process while developing and implementing EST. Studies indicate that EST has contributed largely towards sustainable development in the recent past and there is a rise of a lot of industries which have been contributing towards developing innovative products which are environment friendly.

These barriers to EST are identified through a systematic review of literature and are detailed as under:

Table 1: Variables Identified from Structured Literature Review with their References

Sr. No.	Barriers	List of References
1.	Lack of Awareness	Rahman *et al.,* 2013; Rahim *et al.,* 2012; Azjen, 1991; Legris, 2003; Mathieson, 1991; Bradner, 2003; Carano and Berson, 2007; Abowd *et al.,* 1999; Zhao *et al.,* 2002
2.	Budget Constraints	Sudhakar Reddy, 2003; J.P. Painuly, 2003; Beck and Martinot, 2004; Pepermans *et al.,* 2005; Hasan *et al.,* 2011
3.	Regulatory Concerns	Ghosh *et al.,* 2001; Mirza and Harijan, 2009; Reddy and Painuly, 2003; Fu and Zhang, 2011; Lema and Lema, 2010
4.	Perceived Ease of Use	Davis, 1989; Venkatesh, 2000; Gefen and Straub, 2000; Saadé and Bahli, 2005; Jahangir and Begum, 2008; Suki and Suki, 2011; Hess *et al.,* 2014; Beck and Martinot, 2004; Mirza and Harijan, 2009; Owen, 2006; Painuly, 2000
5.	Perceived Usefulness	Davis, 1989; Segars and Grover, 1993; Saadé and Bahli, 2005; Horst *et al.,* 2007; Saeed and Abdinnour-Helm, 2008; Liaw and Huang, 2013; Johnson, 1994; Painuly, 2000
6.	Low Level of Subsidies	Poh and Kong, 2002; Mirza and Harijan, 2009; Painuly, 2000
7.	Issues in Implementation	Beck and Martinot, 2004; Owen, 2006; Mirza and Harijan, 2009; Junfeng L. *et al.,* 2002
8.	Uncertain Future Legislation	Ghosh D. *et al.,* 2001, Mirza and Harijan, 2009; Outka, 2012
9.	Lack of Financial Support from Government	Mirza and Harijan, 2009, Beck and Martinot, 2004
10.	Lack of Standardization and Uniformity	Painully, 2000

2.1 Lack of Awareness

The rate of technology diffusion is related to the degree of awareness amongst the respondents (Rahman *et al.,* 2013; Rahim *et al.,* 2012). The earlier model focusing on technology adoption, be it theory of planned behavior (Azjen, 1991) or technology acceptance model (Legris, 2003; Mathieson, 1991) focused on awareness as one of the key enablers in the successful diffusion of any technology. Awareness is also considered as a tool for spreading the right set of information to right set of audience. It is worthwhile to look at the initiatives by various agencies to spread awareness in the rural areas (Bradner, 2003; Carano and Berson, 2007; Abowd *et al.,* 1999) ranging from posters, street plays, aganwadi programs for women etc. Lack of awareness usually sabotages the progress and rate of diffusion of any technology (Zhao *et al.,* 2002). The barrier is also discussed in most of the forums by experts from disciplines of environmental economics and technology scientists.

2.2 Budget Constraints

It might be construed that adoption of renewable energy will be cost effective the long run, as there will be a drastic fall in the costs attached to fuel and, hence lower operating costs over a period of time (Beck and Martinot, 2004; Reddy and Painuly, 2003; Figliozzi *et al.,* 2011) however higher initial investment costs can translate into lesser capacity assembled for every Rupee of investment vis-à-vis an investment in the conventional energy sources (Beck and Martinot, 2004). This leads to a consumer bias, as consumers would give preference to containing the initial costs than optimizing the operating costs in the long run (Reddy and Painuly, 2003). In case of India, where majority of the rural consumers fall in the low-income category, cost preference is bound to play a role in their decisions as there is a paucity of cash or credit. The limitation of regular inflow of cash is also found one of the important and well discussed constraints for users in rural areas (Pepermans *et al.,* 2005; Hasan *et al.,* 2011).

2.3 Regulatory Concerns

In India, majority of the RETs (Renewable Energy Technologies are in the nascent stage. The Indian government has not taken any significant steps to ensure proliferation of RETs and has not implemented policies or schemes, which would incentivize the use of RETs by the Indian Industries (Reddy and Painuly, 2003). Also, due to an absence of a structured policy for the participation of the private sector, there is a dearth of participation from the private sector in renewable energy projects (Ghosh *et al.,* 2001; Mirza and Harijan, 2009). Researches also reflect that government participation to frame and implement policies for green technologies have been a weak area and needs attention. The government although have launched some projects, but the impact of these projects on the masses and the rate of diffusion of these technologies has been a serious challenge (Fu and Zhang, 2011; Lema and Lema, 2010).

2.4 Perceived Ease of Use

The terminology perceived ease of use is derived from the early model of Davis (1989). The variable is a key element derived from Technology Acceptance Model. The construct is applicable to any technology and there are multiple studies which uses this construct for their research (Venkatesh, 2000; Gefen and Straub, 2000; Saadé and Bahli, 2005; Jahangir and Begum, 2008; Suki and Suki, 2011; Hess *et al.,* 2014). The renewable energy sector in India has a shortage of skilled and experienced workers, who are exposed to the working and nuances of RETs (Painuly, 2000; Beck and Martinot, 2004). The non-availability of training facilities and infrastructure for enhancement of skills pertinent to RETs, leads to investors in perceiving the RETs as tricky, even though RETs in the long-run are to be cost effective. It is also observed that the perception of users especially in the rural areas is dependent on their familiarity with the technology. Researchers have proved that in a rural set up, the perception is far different from that in the urban areas due to limited interaction with the technologies. (Beck and Martinot, 2004; Mirza and Harijan, 2009, Owen, 2006). Therefore, it is important to analyze the barrier in this study as well.

2.5 Perceived Usefulness

This variable is also derived from Technology Acceptance Model in a seminal study conducted by Davis (1989). The degree of usefulness of the technology needs to be gauged which has come out to be a challenge in studies conducted in the past (Segars and Grover, 1993; Saadé and Bahli, 2005; Horst *et al.,* 2007; Saeed and Abdinnour-Helm, 2008; Liaw and Huang, 2013). The degree of familiarity also leads to perception of users towards environmentally sound technologies as discussed in the previous sections which are backed up by multiple theories and researches. The studies also highlight the fact that people living in rural areas do not see a direct impact of these technologies on their daily functioning and the conventional technologies were not replaced by green technologies (Johnson, 1994; Painuly, 2000).

2.6 Low Level of Subsidies

Studies have indicated that the Indian government has failed to take any steps to ensure use and adoption of EST especially in the rural areas. The firms and the government officials aim at distribution of such technologies one time at a subsidized rate. The functionality of the subsidy is never repeated again for them which reduces their level of motivation to spend as much as they used to buy a conventional technology. Special case studies on green technologies have also discussed the cost of these technologies are higher than the money invested to buy a conventional technology. This makes a rural customer disinterested to choose such technologies again and prefers the conventional technologies over them. On the other hand, subsidies to conventional sources of energy and competing technologies has further deterred the use of the environment sound technologies by giving the competing and conventional sources of energy an artificial advantage (Poh and Kong, 2002; Mirza and Harijan, 2009). In order to stimulate the use of EST, the Indian government needs to ensure that there are provisions of subsidies for EST especially in the rural segment (Painuly, 2000).

2.7 Issues in Implementing

One of the major factors that leads to a problem in the implementation of EST is a shortage of credit faced by investors for renewable energy projects, lack of access to credit stems from various reasons like non-availability of a collateral, unfavorable capital market conditions, poor credit-history, etc. (Beck and Martinot, 2004; Owen, 2006). Another reason that poorly affects the implementation of EST is the absence of coordination and networking with local organizations and community, resulting in a contraction of the flow of credit to the local community which prevents the implementation of EST. (Mirza and Harijan, 2009; Junfeng *et al.,* 2002)

2.8 Uncertain Future Legislation

In order to enable ESTs or RETs to grow and stimulate the use and adoption of ESTs, it is an imperative for the Indian Government to form a central body so as to monitor the overall activities of the energy sector, absence of which will lead to duplication of research and development activities (Ghosh *et al.,* 2001, Mirza and Harijan, 2009). The government should ensure coordination among all ministries and stakeholders when it comes to energy related issues (Outka, 2012; Mirza and Harijan, 2009). It is important for the developers to know the impact and usefulness of the technologies which would help them reframe the strategies and customize their product for different segments.

2.9 Lack of Financial Support from Government

Finance availability to renewable energy projects is restricted due to scant insights into technology, and the resulting high risk perception. Small scale projects fare worse. Further, the availability of finance is often limited to the setting up stage, and does not cover costs of operation (Mirza and Harijan, 2009). In rural areas, the problem is more stark because of unavailability of capital, low reliability of debtors and imperfect credit markets. (Beck and Martinot, 2004)

2.10 Lack of Standardization and Uniformity

Studies have asserted that there is a dearth of facilities for testing the standards of the technologies offered under the category of Environmentally Sound Technology (Painully, 2000). As a result, there is no benchmark in product quality. What plagues this sector the most perhaps is the lack of skill, and compounding that problem is the lack of infrastructure to develop that skill. Till that lacuna can be filled, there will exist a deficit of standard products in the market.

3. INTERPRETIVE STRUCTURAL MODELING— A METHODOLOGICAL PERSPECTIVE

Interpretive Structural Modeling (ISM) enables the individual or a group of them to manage the interrelations between two or more elements at a time without compromising and deviating from the actual properties of the original elements/issues (Morgado *et al.,* 1999).

Term "interpretive structural modeling" (ISM) connotes systematic application of elementary notions of graph theory in such a way that theoretical, conceptual, and computation leverage is exploited to efficiently construct a directed graph, or network representation, of the complex pattern of a contextual relationship among a set of elements (Malone, 1975). ISM is much more flexible than many conventional quantitative modeling approaches that require variables to be measured on ratio scales. It offers a qualitative modeling language for structuring complexity and thinking on an issue by building an agreed structural model (Morgado *et al.,* 1999).

ISM as a tool is interpretive because it is based on interpretation and judgment of group members on whether and how elements are related and it is structural as it extracts overall hierarchy form a complex set of variables. It has a mathematical foundation, philosophical basis and a conceptual and analytical structure. It provides the means to transform unclear and poorly articulated mental hierarchies into visible, well-defined models for better planning of strategies (Barve *et al.,* 2007). Unlike a conventional questionnaire requiring respondents to merely rate the importance of key issues, Interpretive Structured Modeling (ISM) forces the managers to consider various linkages among key issues (Morgado *et al.,* 1999).

ISM allows handling of several elemental classes under various structural types and varied relationships amongst those elements. It helps in understanding of several ill-defined elements that are related in systems (Morgado *et al.,* 1999). It also helps in summarizing relationships among specific items and imposing an order and direction on the complex relationship among elements of the system (Thakkar *et al.,* 2007).

Details of various steps involved in ISM are as follows

1. Identify and list elements/variables relevant to the problem under consideration, through a literature review, field survey or any group activity for the purpose.
2. Use expert opinion or group techniques to determine contextual relationships amongst identified variables, in line with the objectives of the study.
3. Develop a Structural Self Interaction Matrix (SSIM) for variables, indicating pair-wise relationships among variables being studied.
4. Convert the SSIM developed into a reachability matrix.
5. Test the reachability matrix for transitivity (if A depends on B and B depends on C, then by principle of transitivity, A depends on C), make modifications to satisfy the transitivity requirements and derive the final reachability matrix.
6. Delineate levels by iterative partitioning of the final reachability matrix.
7. Translate the relationships of reachability matrix into a diagraph and convert it into an ISM (Interpretive Structural Model).
8. Review the model for conceptual inconsistencies and make modifications in SSIM if necessary.
9. Use the driving power and dependency of each influencer to map the driver-dependency grid for better insight into interdependencies.

Source: Warfield 1974, 1994, 1999; Mandal and Deskmukh 1994; Jharkharia and Shankar 2004; Singh and Sushil 2013; Sharma *et al.,* 2011; Kannan and Haq 2007). The following section will elaborate the steps as mentioned above in detail.

3.1 Structural Self-Interaction Matrix

The structured literature review derived 10 key barriers from the existing literature. The next step is to consult the experts from both academia and industry who are linked with research and innovation in the field of environmentally sound technology. The consultants also include policy makers and people who are members of statutory bodies of research. Table 2 denotes SSIM and uses 4 symbols to denote the relationship between the variables identified from the literature review. These symbols indicate the degree of association between the pairs of the variables and are denoted by 'i' and 'j' (referring to serial number of a barrier in row and column respectively).

V – barrier 'i' needs to be addressed before barrier 'j'
A – barrier 'j' needs to be addressed before barrier 'i'
X – both barriers 'i' and 'j' need to be addressed simultaneously and
O – barriers 'i' and 'j' can be addressed independent of each other

The SSIM matrix denoted by V, A, X and O are represented in the table as under. The table lists the 10 variables vertically and the order of the same is reversed in the horizontal column. This helps in designing a upper triangular matrix with a diagonal running from the top right corner to the bottom left corner. The other half of the matrix is intentionally left blank as the results are always represented from i to j and not the vice-versa.

The next step to complete the remaining half of the matrix and this operation is performed by converting the letter notations to binary values where each of the four notations: V, A, X and O will represent some combination of 1 and 0 in some specific order. The substitution of the rows and columns are done using a specific order and it is as follows:

- If entry (i, j) in SSIM = 'V', enter element (i, j) as '1' and (j. i) as '0' in initial reachability matrix
- If entry (i, j) in SSIM = 'A', enter element (i, j) as '0' and (j. i) as '1' in initial reachability matrix
- If entry (i, j) in SSIM = 'X', enter element (i, j) as '1' and (j. i) as '1' in initial reachability matrix
- If entry (i, j) in SSIM = 'O', enter element (i, j) as '0' and (j. i) as '0' in initial reachability matrix

Table 2: Structured Self-interaction Matrix for the Variables

Sr. No.	Brief Description of Barrier	10	9	8	7	6	5	4	3	2
1.	Lack of Awareness	V	V	V	V	V	V	V	V	X
2.	Budget Constraints	V	V	V	V	V	V	V	V	
3.	Regulatory Concerns	O	O	O	O	O	A	O		
4.	Perceived Ease of Use	O	O	O	O	X	O			
5.	Perceived Usefulness	O	A	A	V	O				
6.	Low level of Subsidies	A	V	V	V					
7.	Issues in Implementation	O	O	O						
8.	Uncertain Future Legislation	A	X							
9.	Lack of Financial Support from Government	O								
10.	Lack of Standardization and Uniformity									

3.2 Transitivity Principle

The completion of the matrix using binary codes makes the reachability matrix ready. As a next step, it is important to also introduce the concept of transitivity to some of the cells of the matrix. A simple explanation to the concept of transitivity may be explained using the following example: If an element 'i' leads to an element 'j' and 'j' leads to an element 'k'; then as per the principle of transitivity 'i' is related to 'k'. This principle of transitivity is used as a basic assumption in interpretive structural model and is used in most of the studies using this technique (Farris and Sage 1975). The objective of the principle is to understand the internal interdependency amongst the variables and apply conceptual consistency thereby aiming towards the robustness of the model. The revised and the final reachability matrix therefore comprises of some entries which are modified with respect to the effect of principle of transitivity.

Final reachability matrix was then obtained for barriers (Table 3) by incorporating the changes necessary to satisfy transitivity requirements detailed in step 5 of Structural modeling methodology. Driving power is defined as total number of variables, which it impacts including itself (equals the count of 1's in a row) and dependency is total number of variables, which have an impact on it including itself (equals the count of 1's in a column).

Table 3: Final Reachability Matrix

Sr. No.	Brief Description of Barrier	1	2	3	4	5	6	7	8	9	10
1.	Lack of Awareness	1	1	1	1	1	1	1	1	1	1
2.	Budget Constraints	1	1	1	1	1	1	1	1	1	1
3.	Regulatory Concerns	0	0	1	0	0	0	0	0	0	0
4.	Perceived Ease of Use	0	0	0	1	0	1	1*	1*	1*	0
5.	Perceived Usefulness	0	0	1	0	1	0	1	0	0	0
6.	Low level of Subsidies	0	0	0	1	1*	1	1	1	1	0
7.	Issues in Implementation	0	0	0	0	0	0	1	0	0	0
8.	Uncertain Future Legislation	0	0	1*	0	1	0	1*	1	1	0
9.	Lack of Financial Support from Government	0	0	1*	0	1	0	1*	1	1	0
10.	Lack of Standardization and Uniformity	0	0	0	1*	0	1	1*	1	1*	1

1* represents transitivity.

3.3 Level Partition and Development of ISM Framework

Final reachability matrix obtained after incorporating transitivity requirements is used for level partitioning. It involves comparing the 'recahability' and 'antecedent' sets of variables and delineating levels on the basis of intersection sets. It leads to a reachability set for a variable by considering the variable itself and other set of variables that causes an impact, whereas antecedent set comprises of the variable and a set of all those variables that have an impact on the primary variable. The hierarchy in ISM is decided by the level of similarity in the reachability and intersection sets. These variables would not impact any other variables.

From the final reachability matrix, the reachability and antecedent set for each antecedent is found (Warfield, 1974). The reachability set consists of the element itself and the other elements which it may help achieve, whereas the antecedent set consists of the element itself and the other elements which may help in achieving it. Thereafter, the intersection of these sets is derived for all the antecedents. The antecedents for which the reachability and the intersection sets are the same occupy the top level in the ISM hierarchy. The top-level element in the hierarchy would not help achieve any other element above its own level. Once the top-level element is identified, it is separated out from the other elements. Then, the same process is repeated to find out the elements in the next level. This process is continued until the level of each element is found (Table 4). These levels help in building the diagraph and the final model. From Table 4 we have developed an ISM model as shown in Figure 1.

Table 4: Level Partition Matrix

Barrier	Level
3, 4	1
5	2
7, 9	3
4, 6	4
10	5
1, 2	6

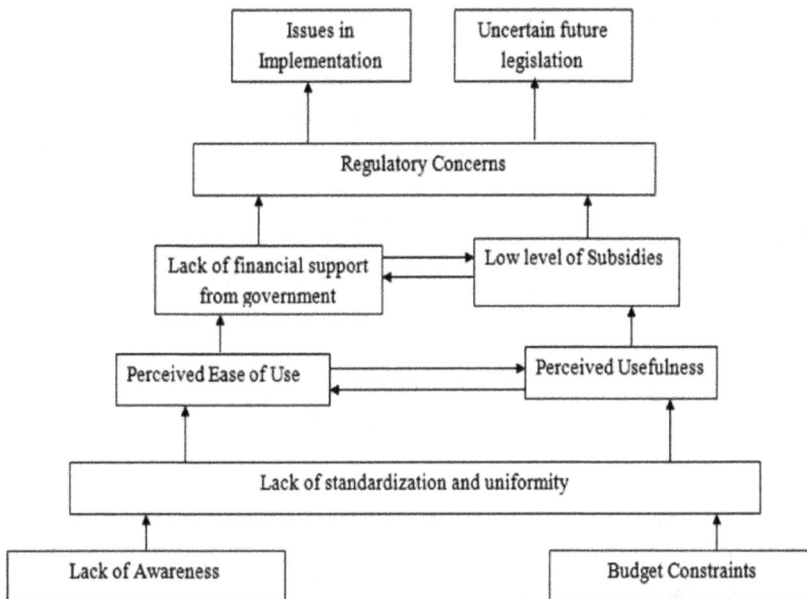

Fig. 1: Hierarchy of Barriers of EST: A Framework Developed Using ISM

4. RESULTS AND DISCUSSION

The developed ISM model consists of six levels of hierarchy as shown in Figure 1. The base level comprises of lack of information and appropriate awareness among public and government agencies and, budget constraints among the users in rural areas. Awareness, although is spread by NGO and other agencies, but due to variation in the intention of the awareness campaigns, the exact message is not delivered to the audience. Also, the agencies involved in this, choose topics which are picked up because of their assignments and research work and, not because there is a need for the discussion. The awareness also lacks teachings regarding to judicious use of money on green technologies, and relating it with the future advantages. This weakens the argument laid by them to enhance the degree of diffusion of technology in the rural areas. The next level of the pyramid is formed by "Lack of standardization and uniformity" of the technologies launched in the rural area. It is observed that there exist a lot of vendors which design different products to be installed in the rural setup. This creates a problem of grievance handling as well. The users face a problem in getting the problem resolved in the malfunctioning of the technology as the people trained to get them resolved are not trained and aware of the different varieties of EST. This makes the user experience unsatisfactory and, therefore adversely affects their perception about the green technologies which can be represented in the third level. The third level of the pyramid as rightly mentioned, deals with the perception of the users that leads to the successful diffusion of any technology. Both the blocks of the third layer are made up of elements like perceived ease of use and perceived usefulness. It is therefore important to simultaneously cater to the concerns of perception as they are interlinked and therefore follow each other. The perceptions of users are not only driven by their internal mindset. It is also found that there are a lot of external barriers apart from the internal barriers which lead to the same. Demography plays an important role in governing the same. Also, it is observed that factors like financial status of the users also influence the perception of users towards usage of EST. However, the same is reflected in the fourth level of pyramid as well. It is worthwhile to understand the target audience of EST in rural areas is not well off in terms of their finances. Thus, the barriers corresponding to the same are interlinked as well. The users are not motivated financially by the government as they do not provide them with subsidies on these technologies and even do not promote the use of EST over the conventional sources of energy. The users are not entitled for any sort of loan for buying EST. Thus, this sector needs a lot of regulation in terms of management and distribution of EST which should cater to their successful diffusion as well. Level five of the ISM framework marks the barrier of government regulation which is impacted dually by the financial constraints and the government support to support the users financially. These moves of the government ultimately lead to a double dip in the rate of diffusion in rural areas. It is also seen that the apex level of the framework is related to uncertainty of the future legislation which corrodes the interests of the users in rural area. As soon as the users achieve a threshold point, their interest starts to dip which leads to discontinuation of the usage of these technologies. It is also worsened by challenges faced by them in installation and maintenance of these technologies. The challenge therefore lies in working on every aspect of the framework to enhance the rate of diffusion of EST in rural India. Special step needs to be taken by the respective stakeholders to instill a sense of responsibility amongst themselves.

5. CONCLUSIONS

This paper includes compilation of 10 barriers of EST which were derived from structured literature review. These variables were used to design a questionnaire which was filled by experts from inter-disciplinary areas. The variables were then structured in a framework using ISM technique and the results of the same are discussed in depth in the paper.

The study for extracting and designing a framework for barriers of EST in rural India helped in pooling insights and resolving debates about the inter-dependency of the barriers. This research could be used to develop and test hypothesis by the research world as it gives relationship between the variables in specific directions. It is also important to understand the pyramid of variables so that action can be taken accordingly in a specific order to cater to the needs of the rural users of the technology. The framework also reveals that awareness of these technologies is found to be one of the major concerns which although lies in the bottom of the pyramid, but is prevalent in the entire humanitarian supply chain. The degree of co-relation between the various stakeholders which leads to the mismatch in the level of diffusion of the technology is also found out to be a concern. Other important concerns, include perception of users towards such technologies and government regulations to enhance the degree of diffusion. Financial constraints and level of subsidies offered by the government and allied bodies were also obstacles which were fitted in the design of the pyramid.

ISM methodology helps in placing these blocks of the pyramid in the most appropriate manner, with practical inputs from the industry practitioners and people from academia as mortar to strengthen the framework. The results also categorizes the barriers in different boundaries with respect to their attributes, thereby giving a broad and a clear picture of the barriers. Although the study tried to give a smooth finish to the framework, yet there are certain limitations which are discussed subsequently.

6. LIMITATIONS OF STUDY

The study uses ISM technique to build the framework which is based on opinions and views of the practitioners and academicians and are interlinked with the literature review. The challenge is always to find out exact variables from structured literature review and pose the right set of questions to the experts. Also, the results which are framed using V, A, X, O are not supported with arguments of the experts. The literature was not specifically found for this broadly coined term EST and specifically for Indian context which might have leave some stones unturned. The study also rests on the shoulders of developing theories from studies conducted in smaller regions of Indian sub-continent and then being generalized which might be true for some variables. Thus, special efforts have been taken to incorporate the minimum degree of errors.

REFERENCES

Abowd, G.D., Dey, A.K., Brown, P.J., Davies, N., Smith, M. and Steggles, P. (1999). "Towards a better understanding of context and context-awareness" in *Handheld and Ubiquitous Computing,* Springer Berlin Heidelberg, pp. 304–307.

Adesina, A.A. and Zinnah, M.M. (1993). "Technology characteristics, farmers" perceptions and adoption decisions: A Tobit model application in Sierra Leone', *Agricultural economics*, Vol. 9, No. 4, pp. 297–311.

Ajzen, I. (1991). "The theory of planned behavior", *Organizational behavior and human decision processes*, Vol. 50, No. 2, pp. 179–211.

Alibeli, M.A. and Johnson, C. (2009). "Environmental concern: a cross national analysis", *Journal of international and cross-cultural studies*, Vol. 3, No. 1, pp. 1–10.

Bai, C. and Sarkis, J. (2010). "Greener supplier development: analytical evaluation using rough Set Theory", *Journal of Cleaner Production*. Vol. 17, No. 2, pp. 255–264.

Balan, S., Vrat, P. and Kumar, P. (2007). "A strategic decision model for the justification of supply chain as a means to improve national development index", *International Journal of Technology Management*, Vol. 40, No. 1, pp. 69–86.

Barve, A., Kanda, A. and Shankar, R. (2007). "Analysis of interaction among the barriers of third party logistics", *International Journal of Agile Systems and Management*, Vol. 2, No. 1, pp. 109–129.

Boltic, Z., Ruzic, N., Jovanovic, M., Savic, M., Jovanovic, J. and Petrovic, S. (2013). "Cleaner production aspects of tablet coating process in pharmaceutical industry: Problem of VOCs emission", *Journal of Cleaner Production*, Vol. 44, pp. 123–132.

Bradner, E., Mark, G. and Hertel, T.D. (2003). "Effects of team size on participation, awareness and technology choice in geographically distributed teams" in *System Sciences, 2003. Proceedings of the 36th Annual Hawaii International Conference*, pp. 10, IEEE.

Carano, K.T. and Berson, M.J. (2007). "Breaking stereotypes: Constructing geographic literacy and cultural awareness through technology", *The Social Studies*, Vol. 98, No. 2, pp. 65–69.

Davis, F.D. (1989). "Perceived usefulness, perceived ease of use and user acceptance of information technology", *MIS quarterly*, pp. 319–340.

Deif, A.M. (2011). "A system model for green manufacturing", *Journal of Cleaner Production*, Vol. 19, No. 14, pp. 1553–1559.

Doss, C.R., 2006, "Analyzing technology adoption using microstudies: Limitations, challenges and opportunities for improvement", *Agricultural Economics*, Vol. 34, No. 3, pp. 207–219.

Dubey, R., Gunasekaran, A. and Ali, S.S. (2015). "Exploring the relationship between leadership, operational practices, institutional pressures and environmental performance: A framework for green supply chain" *International Journal of Production Economics*, Vol. 160, pp. 120–132.

Dües, C.M., Tan, K.H. and Lim, M. (2013). "Green as the new lean: how to use lean practices as a catalyst to greening your supply chain", *Journal of Cleaner Production*. Vol. 40, pp. 93–100.

Farris, D.R. and Sage, A.P. (1975). "On the use of interpretive structural modelling for worth assessment", *Computers and Electrical Engineering*, Vol. 2, No. 2/3, pp. 149–174.

Figliozzi, M., Boudart, J. and Feng, W., 2011, "Economic and environmental optimization of vehicle fleets: impact of policy, market, utilization and technological factors. Transportation research record", *Journal of the transportation research board*, Vol. 2252, pp. 1–6.

Fischer, D. (2012). "Challenges of low carbon technology diffusion: insights from shifts in China's photovoltaic industry development", *Innovation and Development*, Vol. 2, No. 1, pp. 131–146.

Fu, X. and Zhang, J. (2011). "Technology transfer, indigenous innovation and leapfrogging in green technology: the solar-PV industry in China and India", *Journal of Chinese Economic and Business Studies*, Vol. 9, No. 4, pp. 329–347.

Gefen, D. and Straub, D.W. (2000). "The relative importance of perceived ease of use in IS adoption: A Study of e-commerce adoption". *Journal of the Association for Information Systems*, Vol. 1, No. 1, pp. 8.

Gunasekaran, A. and Gallear, D. (2012). "Special Issue on Sustainable development of manufacturing and services", *International Journal of Production Economics*, Vol. 140, No. 1, pp. 1–6.

Hasan, Z., Boostanimehr, H. and Bhargava, V.K. (2011). "Green cellular networks: A survey, some research issues and challenges", *Communications Surveys & Tutorials, IEEE*, Vol. 13, No. 4, pp. 524–540.

Hess, T.J., McNab, A.L. and Basoglu, K.A. (2014). "Reliability generalization of perceived ease of use, perceived usefulness and behavioral intentions" *Mis Quarterly*, Vol. 38, No. 1, pp. 1–28.

Horst, M., Kuttschreuter, M. and Gutteling, J.M. (2007). "Perceived usefulness, personal experiences, risk perception and trust as determinants of adoption of e-government services in The Netherlands", *Computers in Human Behavior*, Vol. 23, No. 4, pp. 1838–1852.

Hsu, C.W. and Hu, A.H. (2009). "Applying hazardous substance management to supplier selection using analytic network process", *Journal of Cleaner Production*, Vol. 17, No. 2, pp. 255–264.

IEA (2007). "Tracking industrial energy efficiency and CO_2-emissions – in support of the G8 plan of action", available at: www.iea.org/textbase/nppdf/free/2007/tracking_emissions.pdf (accessed July 1, 2011)

Jahangir, N. and Begum, N. (2008). "The role of perceived usefulness, perceived ease of use, security and privacy and customer attitude to engender customer adaptation in the context of electronic banking", *African Journal of Business Management*, Vol. 2, No. 1, pp. 32–40.

Jharkharia, S. and Shankar, R. (2004). "IT enablement of supply chains: Modelling the enablers. International", *Journal of Product Performance Management*, Vol. 53, No. 8, pp. 700–712.

Kannan, G. and Haq, N.A. (2007). "Analysis of interactions of criteria and sub-criteria for the selection of supplier in the built-in-order supply chain environment", *International Journal of Production Research*, Vol. 45, No. 17, pp. 3831–3852.

Khanna, M., Isik, M. and Zilberman, D. (2002). "Cost effectiveness of alternative green payment policies for conservation technology adoption with heterogeneous land quality", *Agricultural economics*, Vol. 27, No. 2, pp. 157–174.

Ku, C.Y., Chang, C.T. and Ho, H.P. (2010). "Global supplier selection using fuzzy analytic hierarchy process and fuzzy goal programming", *Qual. Quant*, Vol. 44, No. 4, pp. 623–640.

Legris, P., Ingham, J. and Collerette, P. (2003). "Why do people use information technology? A critical review of the technology acceptance model", *Information & management*, Vol. 40, No. 3, pp. 191–204.

Lema, R. and Lema, A. (2010). "Whither technology transfer? The rise of China and India in green technology sectors", Source: http://umconference. um. edu. my/upload/43-1/papers/262% 20RasmusLema_AdrianLema. pdf.

Lema, R. and Lema, A. (2012). "Technology transfer? The rise of China and India in green technology sectors", *Innovation and Development*, Vol. 2, No. 1, pp. 23–44.

Liaw, S.S. and Huang, H.M. (2013). "Perceived satisfaction, perceived usefulness and interactive learning environments as predictors to self-regulation in e-learning environments", *Computers & Education*, Vol. 60, No. 1, pp. 14–24.

Lo, C.K., Yeung, A.C. and Cheng, T.C.E. (2012). "The impact of environmental management systems on financial performance in fashion and textiles industries", *International Journal of Production Economics*, Vol. 135, No. 2, pp. 561–567.

Luken, R. and Van Rompaey, F. (2008). "Drivers for and barriers to environmentally sound technology adoption by manufacturing plants in nine developing countries", *Journal of Cleaner Production*, Vol. 16, No. 1, pp. S67–S77.

Malone, D.W. (1975). "An introduction to the application of interpretive structural modeling", *Proceedings of the IEEE*, Vol. 63, No. 3, pp. 397–404.

Mandal, A. and Deskmukh, S.G. (1994). "Vendor selection using interpretive structural modelling (ISM)", *International Journal of Operations and Production Management*, Vol. 14, No. 6, pp. 52–59.

Mathieson, K. (1991). "Predicting user intentions: comparing the technology acceptance model with the theory of planned behavior", *Information systems research*, Vol. 2, No. 3, pp. 173–191.

Morgado, E.M., Reinhard, N. and Watson, R.T. (1999). "Adding value to key issues research through Q-sorts and interpretive structured modeling", *Communications of the AIS*, Vol. 1 (1es), pp. 3.

Munshi, K. (2004). "Social learning in a heterogeneous population: technology diffusion in the Indian Green Revolution", *Journal of development Economics*, Vol. 73, No. 1, pp. 185–213.

Pepermans, G., Driesen, J., Haeseldonckx, D., Belmans, R. and D'haeseleer, W. (2005). "Distributed generation: definition, benefits and issues", *Energy policy*, Vol. 33, No. 6, pp. 787–798.

Rahbar, E. and Abdul Wahid, N. (2011). "Investigation of green marketing tools' effect on consumers' purchase behavior", *Business Strategy Series*, Vol. 12, No. 2, pp. 73–83.

Rahim, M.H.A., Zukni, R.Z.J.A., Ahmad, F. and Lyndon, N. (2012). "Green advertising and environmentally responsible consumer behavior: The level of awareness and perception of Malaysian youth", *Asian Social Science*, Vol. 8, No. 5, pp. 46.

Rahman, S.R.A., Ahmad, H. and Rosley, M.S.F. (2013). "Green roof: Its awareness among professionals and potential in Malaysian market", *Procedia-Social and Behavioral Sciences*, Vol. 85, pp. 443–453.

Rao, K.U. and Kishore, V.V.N. (2010). "A review of technology diffusion models with special reference to renewable energy technologies", *Renewable and Sustainable Energy Reviews*, Vol. 14, No. 3, pp. 1070–1078.

Saadé, R. and Bahli, B. (2005). "The impact of cognitive absorption on perceived usefulness and perceived ease of use in on-line learning: an extension of the technology acceptance model", *Information & management*, Vol. 42, No. 2, pp. 317–327.

Saeed, K.A. and Abdinnour-Helm, S. (2008). "Examining the effects of information system characteristics and perceived usefulness on post adoption usage of information systems", *Information & Management*, Vol. 45, No. 6, pp. 376–386.

Sangwan, K.S. (2011). "Development of a multi criteria decision model for justification of green manufacturing systems", *International Journal of Green Economics*, Vol. 5, No. 3, pp. 285–305.

Sarkis, J., Zhu, Q. and Lai, K.H. (2011). "An organizational theoretic review of green supply chain management literature", *International Journal of Production Economics*, Vol. 130, No. 1, pp. 1–15.

Schoenherr, T. (2012). "The role of environmental management in sustainable business development: a multi-country investigation", *International Journal of Production Economics*, Vol. 140, No. 1, pp. 116–128.

Segars, A.H. and Grover, V. (1993). "Re-examining perceived ease of use and usefulness: A confirmatory factor analysis", *MIS quarterly*, pp. 517–525.

Sharma, S.K., Panda, B.N., Mahapatra, S.S. and Sahu, S. (2011). "Analysis of barriers for reverse logistics: An Indian perspective", *International Journal of Model Optimisation*, Vol. 1, No. 2, pp. 101–106.

Singh, A.K. and Sushil (2013). "Modeling enablers of TQM to improve airline performance", *International Journal of Productivity and Performance Management*, Vol. 62, No. 3, pp. 250–275.

Singh, P.J. (2010). "Development of performance measures for environmentally conscious manufacturing", *Ph.D. thesis*, Punjab University, Chandigarh.

Suki, N.M. and Suki, N.M. (2011). "Exploring the relationship between perceived usefulness, perceived ease of use, perceived enjoyment, attitude and subscribers" intention towards using 3G mobile services", *Journal of Information Technology Management*, Vol. 22, No. 1, pp. 1–7.

Testa, F. and Iraldo, F. (2010). "Shadows and lights of GSCM (Green Supply Chain Management): determinants and effects of these practices based on a multinational study", *Journal of Cleaner Production,* Vol. 18, No. 10, pp. 953–962.

Thakkar, J., Kanda, A. and Deshmukh, S.G. (2008). "Interpretive structural modeling (ISM) of IT-enablers for Indian manufacturing SMEs", *Information Management & Computer Security*, Vol. 16, No. 2, pp. 113–136.

Tigabu, A.D., Berkhout, F. and van Beukering, P. (2015). "Technology innovation systems and technology diffusion: Adoption of bio-digestion in an emerging innovation system in Rwanda", *Technological Forecasting and Social Change*, Vol. 90, pp. 318–330.

Tranfield, D.R., Denyer, D. and Smart, P. (2003). "Towards a methodology for developing evidence-informed management knowledge by means of systematic review", *British journal of management*, Vol. 14, No. 207–222.

UNEP, International Environmental Technology Centre. 2008, "Every drop counts: environmentally sound technologies for urban and domestic water use efficiency", *UNEP/Earthprint.*

Vachon, S. and Klassen, R.D. (2006). "Green project partnership in the supply chain: the case of the package printing industry", *Journal of Cleaner Production*, Vol. 14, No. 6, pp. 661–671.

Venkatesh, V. (2000). "Determinants of perceived ease of use: Integrating control, intrinsic motivation and emotion into the technology acceptance model", *Information systems research*, Vol. 11, No. 4, pp. 342–365.

Warfield, J.N. (1974). "Structuring complex systems. Battelle monograph", Vol. 4. Columbus, OH: Battelle Memorial Institute.

Warfield, J.N. (1994). "A science of generic design: Managing complexity through systems design", Iowa: Iowa State University Press.

Warfield, J.N. (1999). "Twenty laws of complexity: Science applicable in organizations", *Systems Research and Behavioral Science*, Vol. 16, No. 1, pp. 3–40.

Young, W., Hwang, K., McDonald, S. and Oates, C.J. (2010). "Sustainable consumption: green consumer behaviour when purchasing products", *Sustainable development*, Vol. 18, No. 1, pp. 20–31.

Zailani, S., Iranmanesh, M., Nikbin, D. and Jumadi, H.B. (2014). "Determinants and environmental outcome of green technology innovation adoption in the transportation industry in Malaysia", *Asian Journal of Technology Innovation*, Vol. 22, No. 2, pp. 286–301.

Zhu, Q. and Sarkis, J. (2004). "Relationships between operational practices and performance among early adopters of green supply chain management practices in Chinese manufacturing enterprises", *Journal of Operations Management*, Vol. 22, No. 3, pp. 265–289.

Non-Lethal Defense Equipment

Aman Jain

Department of Mechanical Engineering, Don Bosco Institute of Technology,
Premier Automobiles, Mumbai–400 070, India
E-mail: rathodaman2293@gmail.com

ABSTRACT: *Killing a human being, whether the person is guilty or innocent often causes a heavy psychological burden. Non-lethal weapons offer an opportunity to save life and to avoid the uncomfortable and sometimes devastating consequences that can accompany the use of lethal weapons. The ideal kind of non-lethal weapon would presumably employ an agent to produce a numbing or other disabling effect; the effect would have to be instantaneous if the individual is in immediate danger. Common non-lethal weapons such as tear gas used in dangerous situations could lead to random weapon firing and an even more serious threat then non action. The best agent would be one that is absorbed immediately into the body through action on the skin although these projectiles would have to penetrate clothing yet not be as dangerous as bullets.*

The goal of the project is to develop non-lethal defense equipment which can be used even by civilians for self-protection. The equipment should have characteristics like easy to handle, small in size, portable, accurate and so on. Hence our aim is to design and construct a working prototype of this kind of a weapon. We have chosen to base our design around the principle of the coil gun which uses the magnetic field generated on passing a current through a coil to propel the projectile to a speed which will achieve the desired result of the project without causing any long lasting injuries. A safe and a healthy society helps in better future of a country. Sustainability towards mankind is one of the sole aim of this project.

Keywords: Non-Lethal, Self-Defense, Coil Gun, Projectiles, Magnetic Field.

1. INTRODUCTION

In the world of magnets and electronics, the coil gun is an interesting and unique application that many people, from casual hobbyists to serious engineers, have tinkered with and built from very simple materials. It operates on the same principles as a solenoid, except in a coil gun, the moving ferrous core inside the solenoid becomes the projectile, and the coil surrounds a longer barrel, instead of just the solenoid housing. By quickly energizing the coil of wire that is wrapped around the barrel, an intense magnetic field is created. This field acts as an electromagnetic slingshot, rapidly attracting the ferrous projectile to the coil, and then the momentum of the projectile itself carries it out of the barrel, and on to the intended target.

For anyone interested in firearms or similar weapons, the idea of a portable, fully automated coil gun presents tremendous potential. Due to the principles that the coil gun operates under, it can launch projectiles of a similar size and weight to conventional weapons, yet there is no loud bang or noise associated with its function, whereas with conventional firearms, you have a sudden, loud noise, and frequently a bright flash. Also with conventional firearms, there is often recoil to deal with, because when the projectile exits the barrel, the high pressure that propelled the bullet down the barrel is released.

The main idea behind a coil-gun is to generate a magnetic field from a solenoid with high currents to attract a ferrous projectile, and then shut down the field and current to let the projectile continue through and out the other end at high velocities.

2. LITERATURE SURVEY

Through research and various papers we have found out the following information.

The basic components of a coilgun include a:

1. power source,
2. a solenoid (the coil),
3. a barrel,
4. a switch (trigger).

These four components will only work together in very low power and low efficiency applications, because with the high voltages and currents encountered in more powerful guns, there are many complications that can arise. These include power switching, EMP and Counter-EMF generation, power loss from resistance, and charging.

Coil-guns currently run at only 1–4% efficiency, but they do not generate large clouds of plasma out the muzzle like rail-guns do at high powers. They also do not suffer from erosion due to electrical arching, and therefore have lower maintenance and operational costs. Coil guns with superconducting materials have reached efficiencies up to 90%.

2.1 Power

The maximum muzzle velocity of a coil-gun is completely dependent on the current that runs through the gun, since the magnetic field generated inside a solenoid is proportional to the current applied and the number of turns divided by the length, as shown in Equation 1,

$$B = (\mu \times n \times I)/L \qquad \qquad \dots (1)$$

Capacitors are the perfect power source because they are able to hold large amounts of energy and release all of it in fractions of a second. The energy a capacitor holds is dependent on the square of the voltage at which charge is stored, shown in Equation 2,

$$U = 0.5 \times C \times V^2 \qquad \qquad \dots (2)$$

A basic capacitor consists of two parallel plates separated by a small distance, as shown in Figure 1.

The capacitance C of a capacitor can be calculated by dividing the amount of charge on each conductor by the voltage between them. It is difficult to have a high voltage capacitor that can hold a lot of charge in a small package, since both increased conductor surface area and increase conductor insulation both require larger internals.

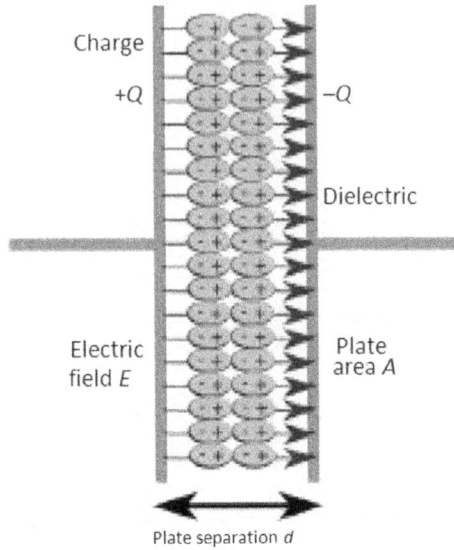

Fig. 1: Basic Capacitor Illustration

In addition to high energy capacity, a good capacitor for use in a coil-gun application must have low ESR and ESL, or equivalent series resistance and inductance. There are also special pulse-rated capacitors that have larger leads and extra heavy duty internals to handle higher currents encountered in pulsed power applications.

2.2 Solenoid (Coils)

The magnetic field generated by a solenoid is given by Equation 1 above, and the generated field shape is shown in Figure 2 below.

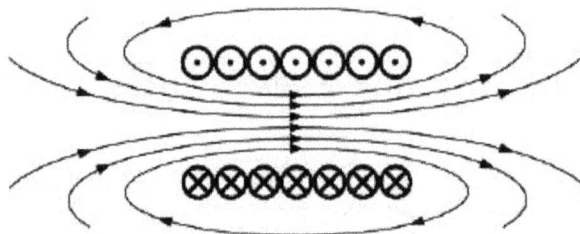

Fig. 2: Solenoid Cross Section with Its Generated Magnetic Field

This cross section of a typical one-layer seven-turn solenoid shows the generated magnetic field when current is running through it. The current is coming out of the page in the bottom and into the page at the top.

The solenoid, along with the power source, are the two most important components because they control how much power can be discharged, and with how much efficiency. The resistance of the solenoid is the main source of resistance the coilgun will have, and is determined by the gauge of wire used. The resistance determines the maximum current pulse that can be generated from the capacitors, since current is voltage divided by resistance, given in Equation 3 below:

$$V = IR \rightarrow I = V/R \qquad \qquad \dots (3)$$

We can calculate the maximum current that will flow through the solenoid with the known resistance of the coil.

Other sources of resistance include the ESR of the capacitors, and the resistance of the power switch, but two are small in comparison with the solenoid. Solenoids, consisting of multiple turns of wire, also have a significant amount of inductance. An inductor resists changes in current, so the more turns a solenoid has, the more inductance it will have, causing the capacitors to have a longer discharge pulse, and lower peak current.

The length of the solenoid itself does not affect the magnetic field it generates as long as the turns per unit length is constant. However, if the solenoid is too short, there will only be a small timeframe to switch off the current before the projectile crosses the midpoint and begins to get pulled back inside, reducing muzzle velocity, or in extreme cases, firing backwards. Having a coil that is too long, in contrast, will reduce the efficiency of the coil-gun. A projectile in the center of a powered solenoid will experience no force, but if moved towards either end, it will get pulled back to the center.

The force is greatest just past the end of a solenoid, and diminishes towards the center. Therefore, even if the capacitor bank was able to provide a high current for the period of time required to pull the projectile from one end into the middle of the solenoid, the projectile is experiencing very little acceleration after entering 1/4 of the solenoid, resulting in wasted energy.

2.3 Power Switching

In order to handle the large amount of current that will flow once the circuit is completed by the switch, special solid-state switches are required. Normal switches and contactors can weld together when trying to complete the circuit, because normal switches do not close instantly. There is a point when the tiny imperfections on the contact plates of a contactor touch before the rest of the plate physically closes together, as well as arching that can occur right before full contact. The resistance of these tiny imperfections can generate immense I^2R losses, which can weld the contacts together before the switch actually fully closes. Hence we use silicon controlled rectifiers (SCRs) in our coil-gun. Silicon controlled rectifiers, also known as thyristors, are made of four layers of alternating P and N type semiconductors. A diagram of the basic SCR is shown below in Figure 3.

By applying a small voltage between gate and cathode, the lower transistor (NPN) will be forced "on" by the resulting base current, which will cause the upper transistor to conduct, which then supplies the lower transistor's base with current so that it no longer needs to be activated by a gate voltage. Once latched "on", there is no way to turn off the SCR until the current flowing through falls below the holding current. This is not a problem in our application, because we want the capacitors to drain fully. If the gate receives a small current pulse but the current through the SCR is less than the holding current, the SCR will not turn on.

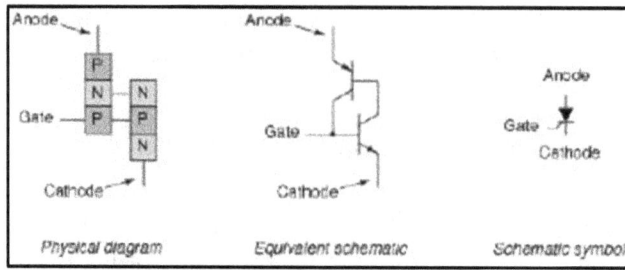

Fig. 3: Diagram and Schematic of an SCR

If used correctly, the SCR is the only logical choice in electromagnetic pulse weapons, with its high pulse current ratings, instant conduction, complete silence, and low (relative to the voltage and current it can pass) trigger pulse requirements.

2.4 Projectile

Kinetic projectiles by definition do not contain any explosive payload, and destroy targets by transferring sheer kinetic energy into shock waves and heat. Projectiles fired by coil-guns and rail-guns fall into this category, since they can be fired at speeds that no conventional weapons can achieve. There are a few special qualities that projectiles used by electromagnetic weapons must have to be effective, in addition to the standard armor piercing abilities. First, the projectile used in a coil-gun must be of ferrous material so that the generated magnetic field can attract the projectile, but it must also be non-magnetizable. If the projectile were to become magnetized, energy would be lost in the magnetization process, and result in lower muzzle velocities.

Second, the higher the magnetic permeability, the stronger the magnetic field can pull on the projectile. However, permeability does not help much if the material saturates too easily. The force a magnetic field exerts on a ferrous material is roughly linearly proportional to the field until the material reaches its saturation point. Past this point, increasing the magnetic field will still increase the force exerted, but with diminishing returns. Therefore, having high permeability and high saturation is required, but if one needed to choose, having high saturation is slightly more advantageous.

In terms of shape, the worst shape is a sphere, since it concentrates the maximum mass into the least amount of volume, thus achieving little acceleration. We chose to use a cylinder shape, since it is easy to make, and fills the barrel completely, eliminating free space between it and the solenoid and allowing maximum magnetic coupling.

As our aim is for a non-lethal weapon we also have to make sure that the projectile design is such that the chance of penetration of skin is the least.

Another important consideration for the projectile is the Ballistic Coefficient (BC).

The formula for calculating the ballistic coefficient for bullets only is as follows:

$$BC_{Bullets} = \frac{SD}{i} = \frac{M}{i - d^2}$$

Where,

BC_{Bullets} = ballistic coefficient

    SD = sectional density, SD = mass of bullet in pounds or kilograms divided by its caliber squared in inches or meters; units are lb/in^2 or kg/m^2.

    i = form factor, $i = \dfrac{C_B}{C_G}$; ($C_G \sim 0.5191$)

    C_B = Drag coefficient of the bullet

    C_G = Drag coefficient of the G1 model bullet

    M = Mass of bullet, 1 b or kg

    d = Diameter of bullet, inches/meters.

This *BC* formula gives the ratio of ballistic efficiency compared to the standard *G1* model projectile. The standard G1 projectile originates from the "*C*" standard reference projectile (a 450 grams (1 lb), 25.4 millimeters (1 in) diameter projectile with a flat base, a length of 76.2 millimeters (3 in), and a 50.8 millimeters (2 in) radius tangential curve for the point) defined by the German steel, ammunition and armaments manufacturer Krupp in 1881. By definition, the G1 model standard projectile has a BC of 1 (using Imperial Units). A bullet with a high BC will travel farther than one with a low BC because it is affected less by air resistance, and retains more of its initial velocity as it flies downrange from the muzzle (see external ballistics).

When hunting with a rifle, a higher BC is desirable for several reasons. A higher BC results in a flatter trajectory for a given distance, which in turn reduces the effect of errors in estimating the distance to the target. This is particularly important when attempting a clean hit on the vital organs of a game animal. If the target animal is closer than estimated, then the bullet will hit higher than expected. Conversely, if the animal is further than estimated the bullet will hit lower than expected. Such a difference from the point of aim can often make the difference between a clean kill and a wounded animal.

This difference in trajectories becomes more critical at longer ranges. For some cartridges, the difference in two bullet designs fired from the same rifle can result in a difference between the two of over 30 centimeters (12 in) at 500 meters (550 yd). The difference in impact energy can also be great because kinetic energy depends on the square of the velocity. A bullet with a high BC arrives at the target faster and with more energy than one with a low BC.

Since the higher BC bullet gets to the target faster, there is also less time for it to be affected by any crosswind.

Sporting bullets, with a caliber d ranging from 4.4 to 12.7 millimeters (0.172 to 0.50 in), have BCs in the range 0.12 to slightly over 1.00 lb/in^2. Those bullets with the higher BCs are the most aerodynamic, and those with low BCs are the least. Very-low-drag bullets with BCs $\geq$ 1.10 can be designed and produced on CNC precision lathes out of mono-metal rods, but they often have to be fired from custom made full bore rifles with special barrels.

Ammunition makers often offer several bullet weights and types for a given cartridge. Heavy for-caliber pointed (spritzer) bullets with a boat tail design have BCs at the higher end of the normal range, whereas lighter bullets with square tails and blunt noses have lower BCs.

Forces and External Factors Acting on the Bullet

In flight, the main force which are acting on a bullet are gravity, drag and if present wind also. Gravity basically imparts a downward acceleration on the projectile and causes it to drop from line of sight. Drag or air resistance decelerates the bullet with a force proportional to the square of the velocity and also cause deviation from its trajectory path.

Long Range Factors

Gyroscopic Drift (Spin Drift): Gyroscopic drift is a spin induced drift experienced by a bullet. A spin stabilized bullet acted by a spin induced sideways component. When bullet rotate in a clockwise direction it experience right side component of drift and vice a versa. This is because the bullet's longitudinal axis and the direction of the velocity of center of gravity deviate by a small angle, which is called equilibrium yaw or yaw of repose.

Magnus Effect: Magnus effect affects the spin stabilized bullet as the spin of bullet creates a force acting either up or down which is perpendicular to the sideways vector of the wind. The Magnus effect induced pressure differences around the bullet cause a downward force when wind is from right and cause a upward force when wind is from left.

Poisson Effect: Drift also occurs due to Poisson effect which depends on the nose of the projectile being above the trajectory. The up tilted nose of the bullet causes an air cushion to build up underneath the bullet and due to that there is increase in friction between this cushion and the bullet, therefore the spin of bullet tend to roll the cushion and move side sideways. Both Poisson and Magnus effect will reverse their direction of drift if the nose fails below the trajectory.

Coriolis Drift: Coriolis drift is caused by the Coriolis and Eötvös effect. These effect cause drift related to the spin of earth. It can be up, down, left or right. It is not an aerodynamic effect.

External Factors

Wind: Wind has a wide range of effects on a bullets. It can deviates the bullet from its line of sight and can causes wind drift called Drag. Drag makes the bullet turn into the wind, and the center of pressure come on its nose which causes a downward deflection. The direction of wind also deflect the trajectory of bullet. A headwind will slightly increase the relative velocity of the bullet, and increase drag and the corresponding drag. Similarly tail wind reduce the drag and the bullet drop.

Vertical Angles: The vertical angle of shot of a bullet will also affect the trajectory of the shot. Gravity on a bullet acts perpendicular to the bullet path. If the angle is up or down, then the perpendicular acceleration will actually be less. The effect of path wise acceleration component will be very small, so shooting up or downhill will both results in similar decrease in bullet drop.

Ambient Air Density: Air temperature, pressure and humidity variations make up the ambient air density. Humidity has a counter intuitive impact. As water vapor has a density of 0.08 gram per liter, while dry air averages about 1.225 grams per liter, higher humidity actually decreases the air density and therefore drag decreases.

2.5 Charging

Charging high-capacity capacitors at high voltages can prove to be quite a problem without spending a lot of money. If money was not an issue, the best way to charge by far would be to use an adjustable industrial DC power supply that can provide a voltage of at least the maximum rating of the capacitors to be charged. Industrial power supplies are by no means cheap, so most home-made coil-guns make use of a variac, step-up transformer, and full wave bridge rectifier in series to charge at an acceptable rate. Another way of charging the capacitors is using something called a boot converter which converts a low DC voltage into a high DC voltage.

A boost converter (step-up converter) is a DC-to-DC power converter with an output voltage greater than its input voltage. It is a class of switched-mode power supply (SMPS) containing at least two semiconductors (a diode and a transistor) and at least one energy storage element, a capacitor, inductor, or the two in combination. Filters made of capacitors (sometimes in combination with inductors) are normally added to the output of the converter to reduce output voltage ripple. Power for the boost converter can come from any suitable DC sources, such as batteries, solar panels, rectifiers and DC generators. A process that changes one DC voltage to a different DC voltage is called DC to DC conversion. A boost converter is a DC to DC converter with an output voltage greater than the source voltage. A boost converter is sometimes called a step-up converter since it "steps up" the source voltage. Since power ($P = VI$) must be conserved, the output current is lower than the source current.

2.6 Efficiency

Since the coil-gun technology is currently 1–4% efficient at best, one of our goals is to maximize the efficiency of my coil-gun. The main losses come from I^2R resistive losses, which primarily come from the solenoid. The only way to reduce resistive losses, then, is to incorporate multiple stages, each with a smaller current pulse. Multiple stages require a much more complicated set-up, involving photo-sensors to properly time the triggering of subsequent coils, but would be well worth the reduction in resistive losses, since the loss is proportional to the current squared.

The other large source of energy losses come from eddy currents generated in all parts of the coil-gun, but most evident in the barrel and projectile. Eddy currents are generated in conductors to oppose the change in flux that created them. The change in flux would be the magnetic field created by the solenoid during firing. The eddy currents circulate around the surface of the conductor they are in, and create their own magnetic field that is opposite to the original field.

2.7 External Iron

The last major improvement to the efficiency of the coil-gun would be the application of external iron. External iron would help with the efficiency of the coil-gun by reducing the size of the magnetic field outside of the solenoid. The iron, if used correctly, will guide the magnetic field through itself from one end of the solenoid to the other end, and will prevent the magnetic field from expanding too far outside the solenoid. This will concentrate more of the magnetic field onto the projectile, thus increasing the force exerted.

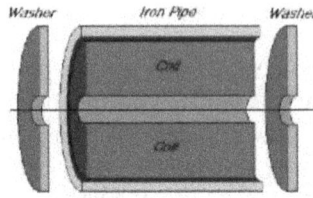

Fig. 4: AutoCAD Rendering of External Iron Placement

The red element is the existing coil, around the empty space inside the barrel. The green elements are iron washers that go on each end of the coil, with and internal diameter equal to that of the barrel. The last element, the light blue element, is the iron pipe that fits around the coil, and is tightly clamped by the two green washers. The thickness of the iron does not affect its ability to redirect magnetic fields because iron has such a high magnetic permeability.

2.8 Potential Problems

The biggest problem that plagues coil-guns involves the very magnetic field that accelerates the projectile. Thousands of amperes of current run though a sufficient sized solenoid will generate strong magnetic fields, but once the current is shut off, the magnetic field will collapse. This changing magnetic field will induce a current in the solenoid which can be equal or greater in magnitude than the original current pulse, albeit shorter. The direction of this current, unfortunately, is in the opposite direction of the original flux that caused it according to Lenz's Law. This will reverse charge the capacitors and overload the SCR, potentially resulting in catastrophic failure for both components. This is known as a back-EMF, or counter-EMF, and can be avoided or blocked with the addition a simple resistor or diode.

The easiest and most effective method of preventing the back-EMF in the first place is to wire in a resistor in series with the solenoid. With the proper resistance, the coil-gun can be made into a critically damped RLC circuit that resists the back-EMF generated in under-damped systems. This method is often not used, however, because a critically damped system will have a very long discharge pulse with a much lower peak current. This will have a detrimental effect on final projectile velocity since the maximum magnetic field generated is much lower than an undamped system. Also, since the solenoid will be energized for an extended period of time, the projectile can pass the center of the solenoid before the discharge finishes, and get sucked back into the solenoid.

Another simple method to prevent back-EMF damage is to direct the current back into the coil over and over, to allow the resistance of the solenoid to dissipate the energy. In order to redirect the current, the use of a flyback diode is required. A flyback diode is basically a diode placed in reverse parallel with the solenoid, as seen in Figure 5 below.

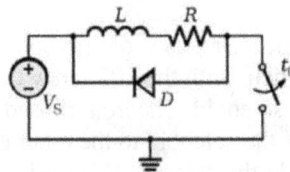

Fig. 5: Diagram of a Flyback Diode in a Simple Inductive Circuit

Vs would be the capacitors, L is the solenoid, D is the flyback diode, R is the resistance of the solenoid, and t0 is the SCR. When firing, the diode blocks current and forces all of the energy through the solenoid until the capacitors are finished discharging. At this instant, the SCR will turn off and create an open point in the circuit. When the magnetic field collapses, it will generate current in the direction of the original pulse. However, since the SCR is now off, the current goes through the only complete circuit: an infinite loop through the diode and solenoid.

2.9 Safety

With a capacitor bank that holds a high amount of energy, any small malfunction could potentially trigger an explosive failure of various components. Care must also be taken to prevent the user from experiencing any electric shocks. A total energy less than 16 joule is thought to be safe in the meaning of not causing fibrillation in a healthy human. It is known that smaller energies have caused death and serious injury because of secondary effects or illness. Violent spasms and shock are some of the dangers. Capacitors are not to be trusted when it comes to rated capacity and voltage. A capacitor that is thought to be safe may have a larger capacity than stated and may be overcharged by a significant amount. Another hazard of capacitors is dielectric absorption. This effect sometimes allows capacitors to recharge themselves without any external power. Any hazardous capacitor should be stored with a short circuit across the terminals in order to prevent this and should always be considered potentially charged otherwise. Due to the various risks, all precautions should be take while handling the components and assembly.

3. PROBLEM STATEMENT AND PROPOSED SOLUTION

There are many situations in life in which people are in danger. In these situations they require something to protect themselves. This can be done by conventional means but at the risk of causing permanent damage.

As evident from our project title, the aim of this project is to create a tool which would allow people to protect themselves without causing serious harm to the individuals which the tool is being used on. Hence our aim is to propel a projectile at speeds that want cause any serious harm but will stun or incapacitate an individual for a short period of time till help/more people arrive.

The way we plan to do this is using the concept of the coil gun which at the current state of technology will be capable of meeting the necessary requirements of our project.

4. DESIGN AND ANALYSIS

4.1 Projectile Velocity

The first parameter that was decided was the projectile velocity. As such you would not say that the projectile velocity was decided but the kinetic energy transferred to the projectile.

Assuming a mass of 10 grams, we decided on a velocity of minimum of 30 m/s. the figure of 30 m/s was decided based on research. In our research we determined the ideal velocity of the projectile to cause maximum effect without penetrating the skin or any other body part of an individual. We found that for a 10 gram projectile, velocity in the range of 25–30 m/s range would not cause any skin penetration. The projectile would have an exit kinetic energy of about 3.125–4.5 joules.

Hence it was decided that we would aim for a final projectile velocity of 30 m/s.

4.2 Projectile

As stated above, he worst shape is a sphere and the best shape is a cylinder. Hence we decided on a cylindrical projectile. The projectile was designed to be as aerodynamic without affecting the aim of the project. There were two iterations of the bullet design. Once the first design was made we felt it would not couple well with the magnetic field created by the coil and hence we decided to redesign the projectile. The projectile was designed taking drag and ballistic coefficient into consideration. Certain features of the projectile were included to reduce drag and turbulence. For example the chamfer given at the end of the projectile is given to reduce the amount of turbulence.

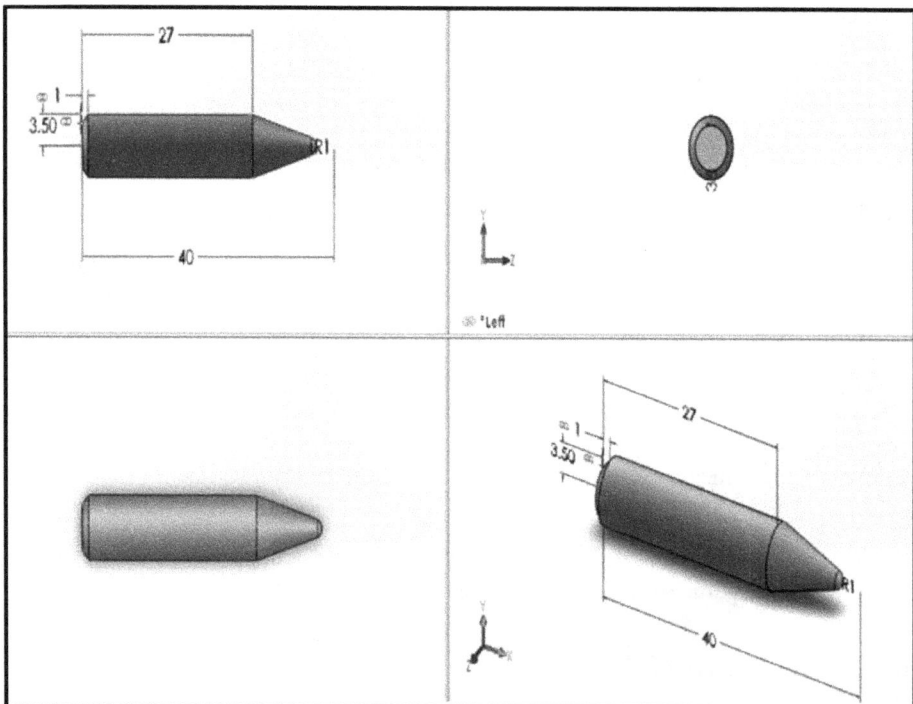

Fig. 6: CAD Model of Final Projectile

Once the CAD model of the projectile was ready we performed a CFD analysis on it. In which we found out the coefficient of drag. The CFD analysis was performed using ANSYS fluent. The geometry was created and meshed using the inbuilt software and then the necessary boundary conditions were inputted and the problem was solved for the design velocity.

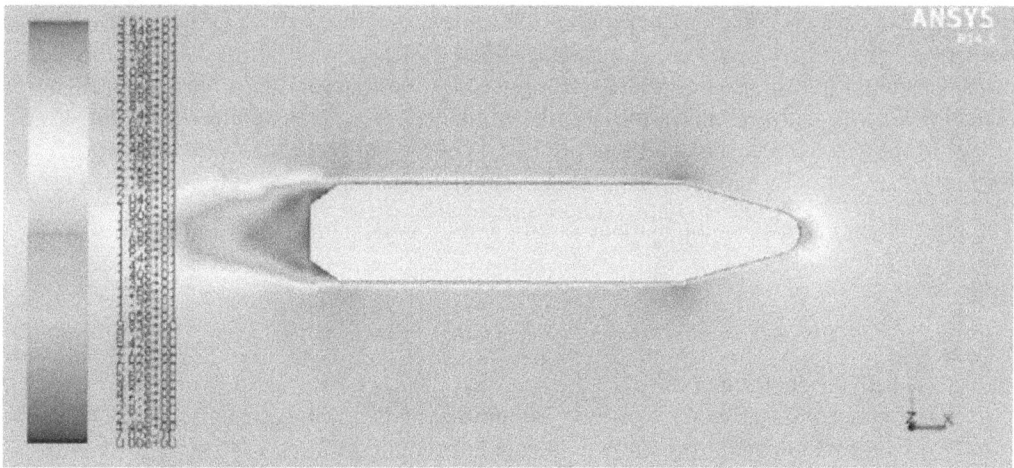

Contours of Velocity Magnitude (m/s)

Mar 27, 2014
ANSYS Fluent 14.5 (3d, pbns, k-kl-w)

Fig. 7: Velocity Contour of Final Projectile Obtained from CFD Analysis

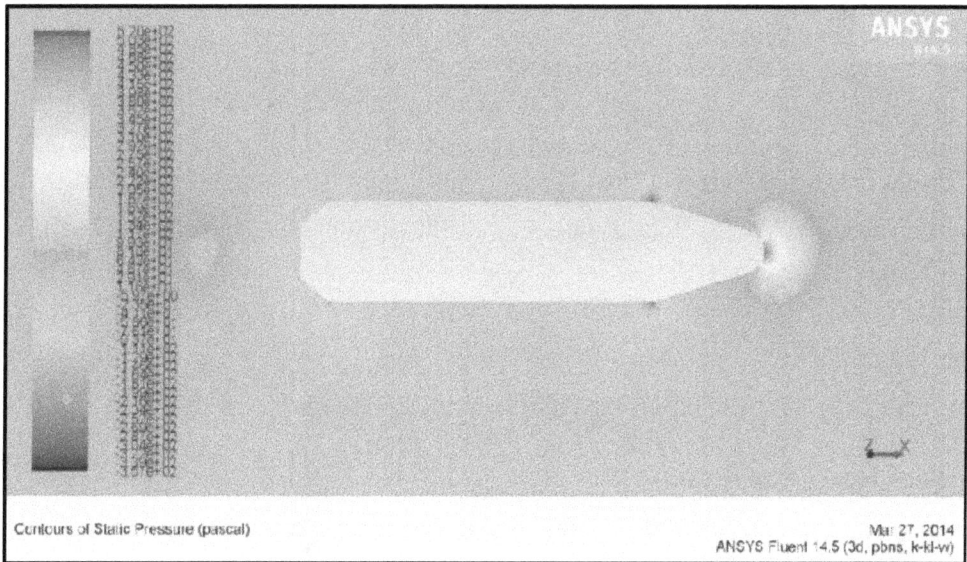

Contours of Static Pressure (pascal)

Mar 27, 2014
ANSYS Fluent 14.5 (3d, pbns, k-kl-w)

Fig. 8: Pressure Contour of Final Projectile Obtained from CFD Analysis

The analysis gave us the value for coefficient of drag (Cd) Cd = 0.290.

The Cd value of the projectile is similar to the Cd value of bullets used in conventional guns hence we find that the value is satisfactory.

Ballistic Coefficient of the Projectile

The ballistic coefficient (BC) of a body is a measure of its ability to overcome air resistance in flight, it is inversely proportional to the negative acceleration–a high number indicates a low negative acceleration. This is roughly the same as saying that the projectile in question possesses low drag, although some meaning is lost in the generalization. BC is a function of mass, diameter, and drag coefficient. It is given by the mass of the object divided by the diameter squared that it presents to the airflow divided by a dimensionless constant "i", that relates to the aerodynamics of its shape. Ballistic coefficient has units of lb/in² or kg/m². BCs for bullets are normally stated in lb/in² by their manufacturers without referring to this unit.

The formula for calculating the ballistic coefficient for bullets only is as follows:

$$BC_{Bullets} = \frac{SD}{i} = \frac{M}{i - d^2}$$

Where:

$BC_{Bullets}$ = ballistic coefficient

$\quad SD$ = sectional density, SD = mass of bullet in pounds or kilograms divided by its caliber squared in inches or meters; units are lb/in² or kg/m².

$\quad i$ = form factor, $i = \dfrac{C_B}{C_G}$; $(C_G \sim 0.5191)$

$\quad C_B$ = Drag coefficient of the bullet

$\quad C_G$ = Drag coefficient of the G1 model bullet

$\quad M$ = Mass of bullet, lb or kg

$\quad d$ = Diameter of bullet, inches/meters

In our case,
- mass of the projectile = 11.37 gms or 0.025066559 lbs
- diameter of bullet = 7.1 mm or 0.279528 in

From our cfd analysis of the projectile, the drag coefficient of the bullet was found to be 0.290

Form factor (i) = 0.290/0.5191
$i = 0.5587$

Therefore the Ballistic Coefficient (BC) is 0.5742.

This again is similar to projectiles used in conventional weapons.

4.3 Selection of Capacitors

The capacitors were chosen based on what was available in the market. It was decided that we would require capacitors of a 300 V rating and hence we searched capacitors which would come close to what we needed. Through rough calculations it was determined we would need capacitors with a capacitance value of 1800 micro farad. We found capacitors which were similar but not exactly what we needed hence we had to make certain changes to our rough calculations.

The specifications of the capacitors are:

Rated Voltage   : 330 Volts
Rated Capacitance : 2000 Micro Farad
Height   : 7 cm
Width   : 3.5 cm

Fig. 9: Image of Capacitor Used

Two capacitors were purchased, one for each stage.

The capacitor is an electrolytic capacitor. An aluminum electrolytic capacitor is just fine for the application we're undertaking since electrolytic capacitors have the second highest.

Capacitance rating below super/ultra-capacitors and can achieve much higher voltages for the tradeoff. Ceramic and film capacitors can achieve higher voltages at lower capacitances, so it is a trade off in design. Aluminum electrolytic in this regard also averaged out to be more cost effective for the given ratings, so everything works out.

4.4 Coils

The coils can be designed only after the projectile and capacitors have been decided. This is due to the fact that they have to discharge all the energy stored in the capacitor before the center of mass of the projectile passes the center of the coil, else the coil will be sucked back which will result in a loss of velocity. Hence any change in projectile or capacitor will result in a change in the coil specifications. As we have two stages and the entry velocity will be different at each stage, each coil will be different.

We use enamel coated copper wires for the coils. These copper wires are used in transformers and electric motors. We use copper wire of size 20 AWG. This is capable of sustaining 149 A for 1s or 832 A for 32 ms. As the current is flowing to the coil only for up to 5 ms, it is acceptable if the peak current is larger than the rated 832 A for 32 ms.

Table 1: Coil Specifications

	Coil for 1ˢᵗ Stage	*Coil for 2ⁿᵈ Stage*
Number of Turns	276	184
Winding Length (cm)	4	4
Internal Diameter (mm)	10	10
External Diameter (mm)	21	17
Wire Length (m)	13.94	8.34
Turns per Layer	46	46
No. of Layers	6	4
Coil Resistance (Ohms)	0.4644	0.2808
Coil Weight (Grams)	64.4	38.9

The coils were designed by trial and error using an RLC simulator.

Using the formula,

$$V_{exit} = ((2 \times \text{Volume} \times \mu_0 \times X_m \times n^2 \times I^2)/\text{Mass})^{1/2}$$

Expected exit velocity of the first stage = 26.84 m/s
Expected exit velocity of the second stage = 34.02 m/s

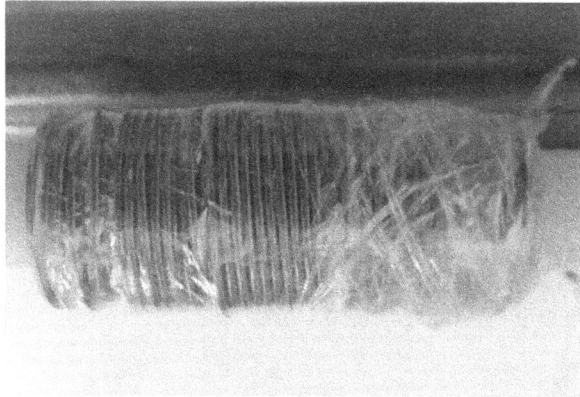

Fig. 10: Image of Coil

The coils were covered by tape to make sure that they are held in place and do not become loose and so that they keep their shape.

One factor to make sure of was that the coils would not over heat as the enamel coating starts to get damaged after a temperature greater than 105°C is reached. After which the enamel will start to crack after prolonged use at this temperature. Hence we check the temperature rise in the coil for every shot fired.

Each coil is subjected to 108.9 J of energy for every shot fired. Say 70% of this energy is lost due to resistance of the coil which results in the increase of coil temperature.

Therefore,

108.9 × 0.7 = Mass of Coil × Cp of Copper × Temperature change

76.23 = 0.0644 × 381 × Temperature change

Temperature Change = 24.5364°C

Hence, assuming an ambient temperature of 27°C, we are nowhere near the temperature limits of the coil wire.

4.5 Selection of SCR

Based on the coil design the estimated peak current flowing through the circuit is found.

Using this peak current value an appropriate SCR is selected.

The peak current value calculated was 906.4 A.

SCRs have a peak current rating which they can take for a few seconds without causing any damage to the SCR. This current rating is usually for 10 sec. hence we choose an SCR which has a peak current rating similar to the calculated peak current.

Hence we selected the NTE5731.

Two SCRs were purchased one for each stage.

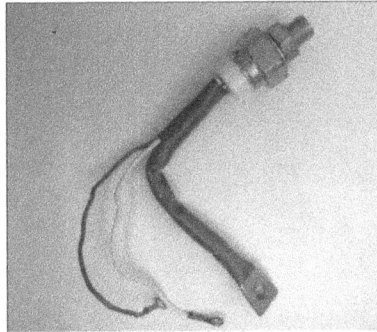

Fig. 11: Image of SCR Used

4.6 External Iron

To make sure that adding external iron to the design was worth the extra weight it added, we did a FEMM analysis on the effect of external iron on the force acting on the projectile. The analysis was done using FEMM v4.2 which is a free to use software available.

The analysis was done using axisymmetric condition along the axis of the coil. A 2D analysis was carried out. Material properties and boundary conditions were inputted into the software as well as the coil specification.

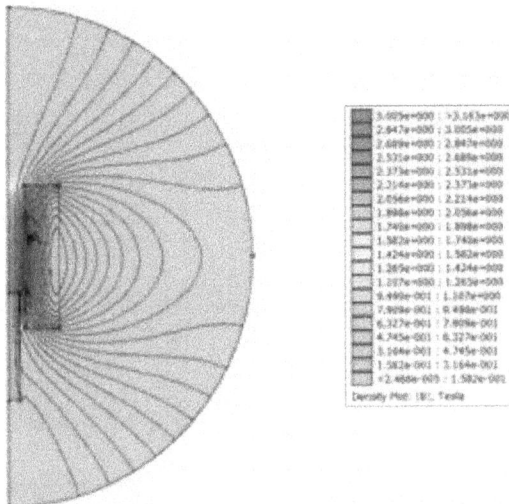

Fig. 12: Magnetic Field Contour of Coil without External Iron

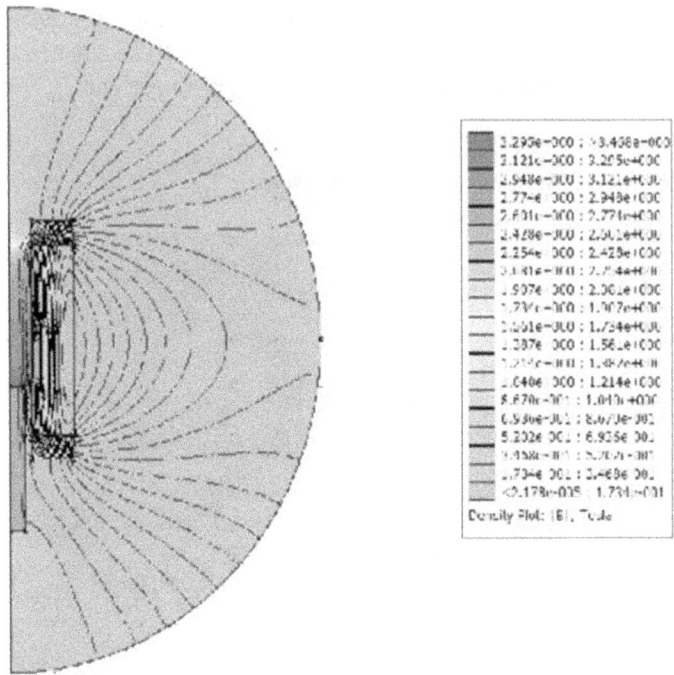

Fig. 13: Magnetic Field Contour of Coil with Washers at the End of the Coil

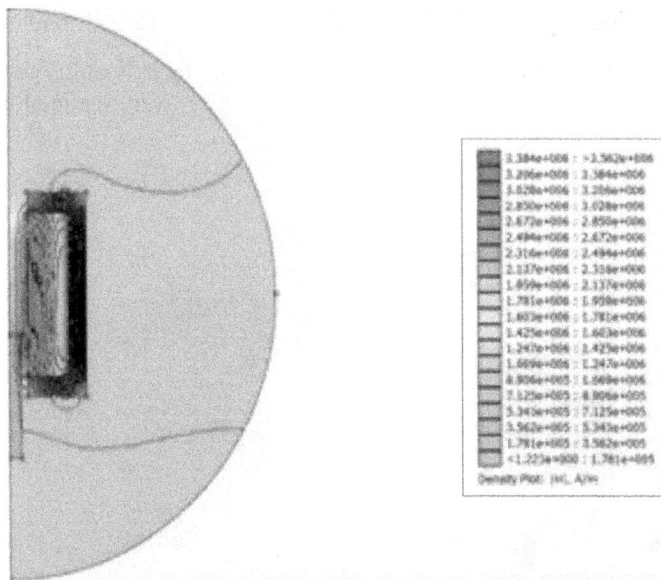

Fig. 14: Magnetic Field Contour of Coil with External Iron

From the analysis we concluded that due to the external iron there was an increase in magnetic field strength which resulted in an increase in the force acting on the projectile.

Table 2: Force Exerted on Projectile for Different Conditions

	Force (N)
Without External Iron	161.512
With Only Washers at the Ends	173.852
With External Iron	194.394
External Iron with 1mm gap between Coil	191.56

It is evident from the above table that the external iron provides an improvement in force and also the smaller the gap between it and the coil the better.

4.7 BARREL

The barrel is made of a non-metallic material as there would be loses due to eddy currents in metallic barrels. The barrel also has to be strong enough to resist the compressive forces that the coils apply on the barrel during operations.

The internal diameter of the barrel has to be large enough for the projectile to pass through with the least amount of resistance. The external diameter has to be as small as possible so that the gap between the coil and the projectile is as small as possible but the diameter should be large enough so that the barrel is strong enough.

Hence we decide on,

Internal Diameter = 8 mm
External Diameter = 10 mm

4.8 Reloading Mechanism

The weapon needs a reloading method just in case more than one projectile has to be fired. Hence we decided to include a way to fire more than a single projectile. We decided to base our design on what you would see on conventional weapons. The mechanism consists of a spring in a casing which houses the projectiles. As a projectile is fired, another one is pushed up by the compressed spring. There are openings at the top so that the spring return solenoid can push the projectile into the barrel.

4.9 Boost Converter

The charging circuit is composed of two parts:

The oscillating circuit and the step up circuit. It gets power from the 11.1 V battery. The purpose of this circuit is to charge the capacitor that provides the energy for the coil.

The oscillating circuit will provide a pulse with a maximum at to the step up circuit a high frequency oscillator will be implemented with an IC555 chip.

The waveform generator will use the 555 timer circuit to generate the required waveforms. The frequency we used is somewhat arbitrary but explained below in the step up circuit.

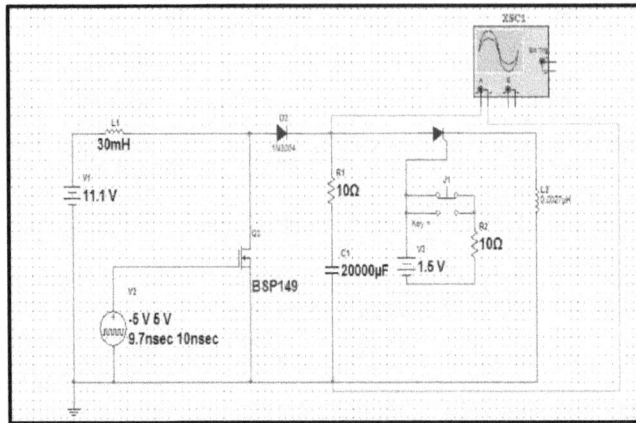

Fig. 15: Schematic of Boost and Oscillating Circuit

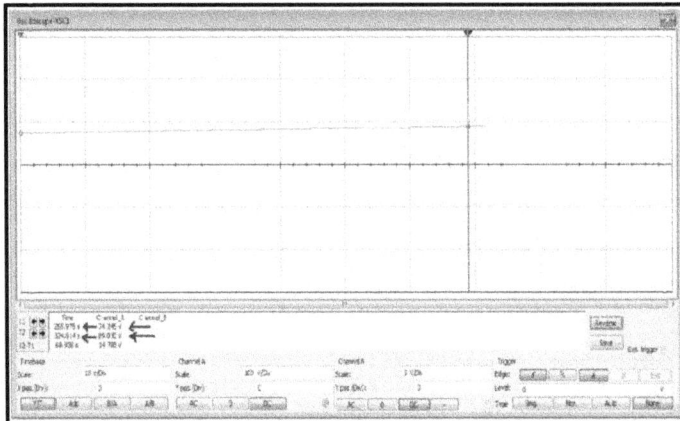

Fig. 16: Output Graph Using Simulation of the Booster Circuit in MULTI-SIM

The step up circuit is a simple boost converter. High powered diodes will be used. These diodes can withstand the 1000 V output. We used a simple boost converter to step up the voltage from 11.1 volts to 250 volts. The equation for a boost converter is

$$V_{in}/V_{out} = 1-D$$

If we want to use a smaller inductance in the boost converter, we use a higher frequency. Once we set the frequency, we varied the inductance to achieve an output voltage of 250 volts from an input voltage of 11.1 volts.

5. MANUFACTURING

5.1 Projectile

The projectile was made on the lathe. It is made of mild steel. The raw material were mild steel cylinders of 5 cm length and 10 mm diameter.

Fig. 17: Image of Initial Projectile

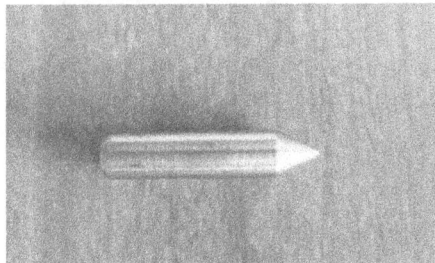

Fig. 18: Image of Final Projectile

5.2 Barrel

The barrel was made by taking a solid plastic pipe cylinder and turning it to the required external diameter and drilling the cylinder for the internal diameter. The raw material dimensions were 15 mm diameter and 20 cm length.

5.3 Body and Reloading

The body and reloading mechanism was made out of plywood. The spring used in the reloading is a coil spring.

Fig. 19: Image of Completed Body

Fig. 20: Image of Magazine for Reloading Mechanism

5.4 External Iron

The external iron was made from pipe of suitable dimensions which was cut and machined so as to achieve the required dimensions.

Fig. 21: Image of External Iron

5.5 Assembly of Coil and Barrel

The coils were wound directly on the barrel and then covered in tape so as to maintain its shape.

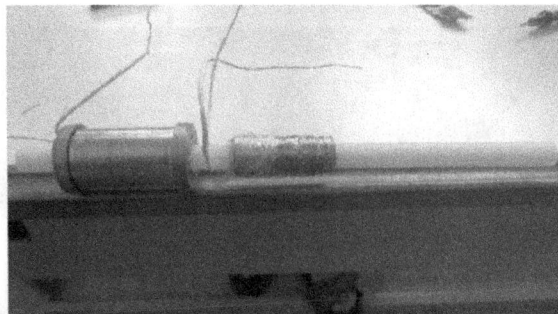

Fig. 22: Image of Assembly of Coil and Barrel

5.6 Boost Converter

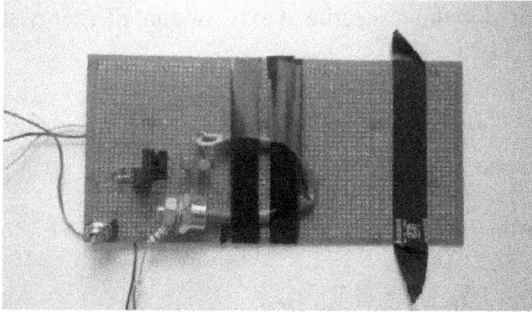

Fig. 23: Image of Boost Converter Circuit

The boost converter was constructed on a circuit board

6. TESTING

First the boost converter was tested using capacitors of smaller capacity.

After this the complete circuit was connected for one stage.

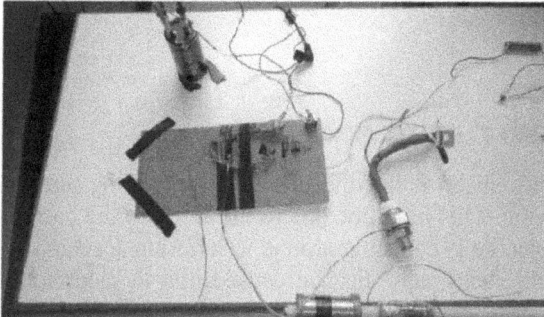

Fig. 24: Image of Circuit during Testing

After the capacitor was charged for a certain period of time, we fired a shot.

The projectile was fired when the capacitor was charged to a voltage of 55 V and at 80 V.

7. RESULT AND DISCUSSION

During the testing the following results were observed:

1st Test

Voltage of capacitor at before test shot = 55 V
Amount of energy stored in the capacitor = $0.5 \times 2000 \times 10^{-6} \times 55^2 = 3.025$ J
Exit velocity of the projectile = 4.8 m/s

Energy stored in the projectile at exit = $0.5 \times 0.01137 \times 4.8^2 = 0.1309824$ J

Efficiency = Energy Stored in the projectile at exit/Amount of energy stored in the capacitor
= 0.1309824/3.025
= 0.043299
= 4.3299%

For a fully charged capacitor:
Using above efficiency, we get
Energy stored in projectile = 3.89699 J
Velocity of projectile = 26.1818 m/s

This is similar to the expected exit velocity of the first stage.

2nd Test

Voltage of capacitor at before test shot = 80 V
Amount of energy stored in the capacitor = $0.5 \times 2000 \times 10^{-6} \times 80^2 = 6.4$ J
Exit velocity of the projectile = 6.0 m/s
Energy stored in the projectile at exit = $0.5 \times 0.01137 \times 6.0^2 = 0.20466$ J

Efficiency = Energy stored in the projectile at exit/Amount of energy stored in the capacitor
= 0.20466/6.4
= 0.03197
= 3.1978%

8. CONCLUSION AND FUTURE SCOPE

From the above test we observe that the efficiency of the system is close to the expected efficiency but it is varying. This might be due to the fact that the barrel internal diameter is too large and hence the projectile is bouncing around in it during acceleration and losing some of its energy. The varying efficiency may also be caused due to incorrect starting position on the projectile.

Another problem encountered is that we are not able to charge the capacitor fully as it is taking a large amount of time to reach maximum capacity. We have been able to charge it only up to 7.1% of its peak capacity. This is due to the boost converter not being able to supply energy to charge the capacitor at the required rate. Hence we have to modify the boost converter or look for an alternative method to charge the capacitor.

Once the above problems are rectified, we have to check whether the desired velocity is reached and whether the second stage is necessary.

We have to conduct tests to determine the exact staring position of the projectile to achieve maximum efficiency.

Even though we know that external iron has an effect to increase the force acting on the projectile. We have to conduct test to measure the amount it increase the efficiency of the system. Hence we will find out whether it will be advantageous to add external iron to the second stage, if necessary.

REFERENCES

Effect of Projectile Design on Coil Gun Performance: Jeff Holzgrafe, Nathan Lintz, Nick Eyre and Jay Patterson.

Coilgun Technology at the Center for Electromechanics: Bresie, D.A. and Ingram, S.K.

Electromagnetic Acceleration: Asynchronous Linear Induction Motors: By Wilson Wong.

FEMM 4.2 Magnetostatic Tutorial: David Meeker.

A Brief Theoretical Study of Reluctance Based Mass Accelerators and Actuators: Bill Slade.

Reliability Data to Improve High Magnetic Field Coil Design for High Velocity Coil guns. Kaye, Ronald J. and Mann, Gregory A.

Induction Coil guns for hyper velocities: Prof. Laithwaite, E.R.

Pulsed Coil gun limits: Stephen Williamson and Alexander Smith.

Simulation and Analysis of Electromagnetic Coil gun: Liuming GUO, Shuhong WANG, Ningning GUO, Jie QIU.

Coil Gun Electromagnetic Launcher (EML) System with Multi-stage Electromagnetic Coils: Su-Jeong Lee, Ji-Hun Kim, Bong Sob Song and Jin Ho Kim.

APPENDIX

Cost Incurred

Component	Cost
Capacitors	1600
SCRs	5000
Spring Return Electromagnet	370
Coils	1000
Wood	250
Boost Converter Components	700
5.5 m Copper Wire	150
Gloves, Soldering Iron, Solder Wire	448
Multi-meter	200
Miscellaneous	1000
Total	**10718**

Experimental and Simulation Study of Modified Acoustic Horn Design for Sonic Soot Cleaning

Arti Vishwanath* and Deshmukh N. Nilaj

Department of Mechanical Engineering, Fr. C. Rodrigues Institute of Technology,
Vashi, Navi Mumbai, India
*E-mail: arti.2905@gmail.com

ABSTRACT: *Ash deposits and dust particles are formed inside process equipments like boilers and electrostatic precipitators; and in material handling equipments like hoppers and silos during its operation. These deposits hinder the working of the equipment, gradually reducing its efficiency. So, removal of these deposits at regular intervals is necessary for smooth operation of the equipments. Sonic soot cleaning is a method which uses sound waves to dislodge the deposits. It employs an acoustic horn which is the source to produce sound and is designed such that it produces a frequency of sound that matches with the natural frequency of the particles to be cleaned.*

In this study, modified geometries of an acoustic horn are designed and fabricated so that the desired Sound Pressure Level (SPL) and frequencies are obtained for effective cleaning. The frequency can be reduced by increasing the horn length. The increase in length however causes installation problems, due to space constraint. To overcome this problem, numbers of bends are introduced to increase the overall length of the horn. Using these modified designs; frequency and sound pressure level are obtained experimentally as well as by conducting simulations for all cases, and results obtained from both are compared. It is seen that for the case where horn is attached to 6, 8 and 10 number of bends; frequency obtained fell in the range of desired frequency. The percentage of error obtained after comparison is found to be within acceptable limits.

Keywords: Sonic Soot Cleaning; Acoustic Horn; Frequency; Sound Pressure Level.

1. INTRODUCTION

Industrial process equipments are delivered clean with no soot, slag and/or scale. Based on different mechanisms of formation of ash deposits[14] on the heat transfer surface, two general types of ash deposition have been defined namely slagging and fouling. If the deposits are soft, loose and slimy it is called as slag or sludge. If the deposits are hard and adhering to the inner walls of surface it is called as fouling or scale. After a while, ash deposits in the form of scales are formed on the inner surfaces of the equipment during its operation. If at regular intervals, these deposits are not cleaned, it will eventually decrease the efficiency of the equipment. Many techniques exists and are used to clean the surface of process equipments[1] viz. internal cleaning, acid cleaning, mechanical cleaning and steam soot blowing. However, these methods had their own limitations mostly of shutting down of the equipment resulting in increased system downtime during cleaning and corrosion problems which led to the invention of sonic soot cleaning. Sonic soot cleaning is a method applied to remove the ash and/or fossil deposits by the use of sound waves and its energy.

2. SONIC SOOT CLEANING

Sonic soot cleaning is a technique which employs compressed air operated blowers. The blowers are designed in such a way that it will produce sound waves at a frequency that matches with the natural frequency of the particles to be cleaned and when the sound waves strike the particles, due to resonance, it vibrates and falls off the surface. L. Buhl[8] concluded that the most correct physical explanation of how acoustic waves dislodge the dust was that gas movement influences the particles due to friction at the surface and thus removes them.

2.1 Advantages and Applications of Sonic Cleaning

Sonic soot cleaning has many benefits[6] *i.e.* improved efficiency of equipments,[9] less energy dissipation, reaches everywhere of the surface due to reflection and diffraction of sound waves. Also, as air is used no surface corrosion problem arises. It is better from economic point of view as it requires less maintenance and operational cost. Sonic soot cleaning finds many proven applications *i.e.* in boilers, boiler tubes, heat exchanger surfaces, bag house fabric filters, electrostatic precipitators,[10] hoppers, silos, evaporative gas cooling systems and spray dryers.

2.2 Construction and Operation of Acoustic Horn

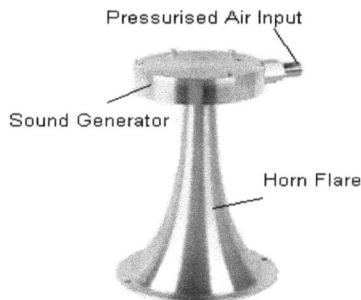

Fig. 1: Acoustic Horn

An acoustic cleaner consists of two main components, a compression driver and a horn as seen in Figure 1.[13] The horn is a component with its gradual change in cross sectional area from throat to mouth that increases the efficiency of sound radiation by matching the acoustic load driven at the horn throat. The narrow part of the horn next to the driver is called as the throat. The larger part at the other end is called as the mouth. Horn flares are used to control the spatial distribution of sound radiating from the horn mouth. An approximate equation can be used to estimate the performance characteristics of horns, provided the function that governs the change in cross sectional area is simple.

From the pressurized air input, compressed air enters the horn and sets the metallic diaphragm into mechanical motion. The movement of the diaphragm produces vibrations resulting in pressure fluctuations. As these pressure fluctuations travel through the horn, sound waves are generated.

3. DESIGN OF MODIFIED HORN GEOMETRY

G. Seiffert[5] studied samples of dust collected from different power plants and boilers. He found that the dust samples collected showed similar characteristics. The size of deposits[8] on an average was found to be 50 microns across their largest dimension. Our basic aim is to obtain desired SPL at a lesser frequency for effective cleaning. From literature survey, it is concluded that the natural frequency of the dust particles lies in the range of 50 Hz–450 Hz. The desired SPL required[2] for efficient cleaning is 140 dB–150 dB. In order to obtain such less frequency, it is necessary that the length of acoustic cleaner should be increased. It is also seen that the overall length[16] of acoustic cleaner lies in the range of 0.4 m–2 m. To avoid space restrictions within process equipments, the overall length of horn can be distributed in form of bends.[17] For this study, six cases are considered for carrying out simulation and experimentation, that is, plain horn and horn attached to 2, 4, 6, 8 and 10 numbers of bends.

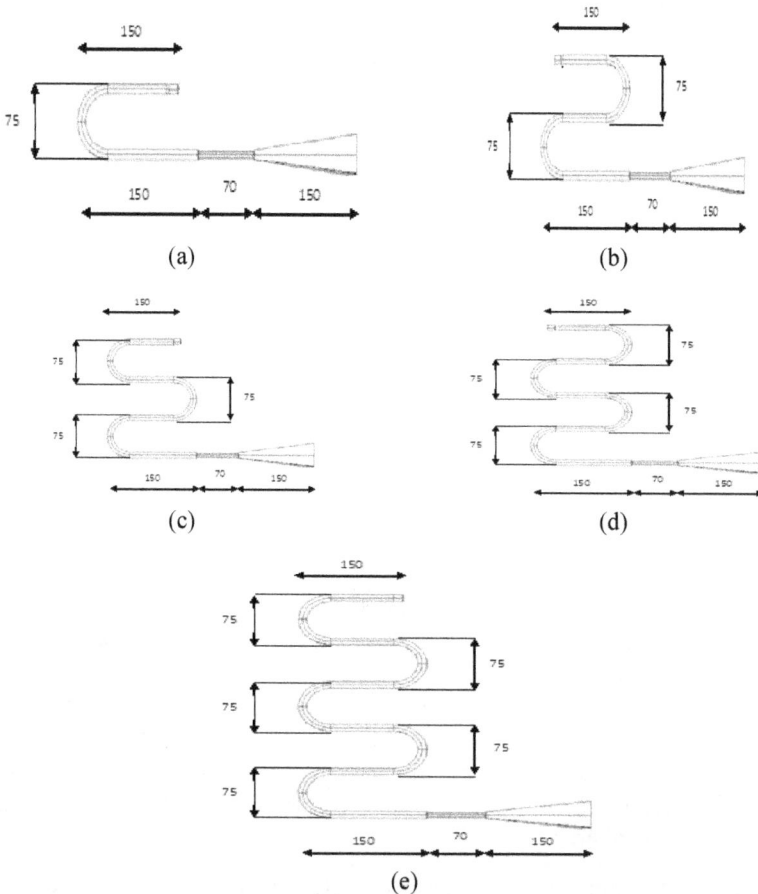

Fig. 2: Dimensions of Horn Attached to (a) 2, (b) 4, (c) 6, (d) 8 and (e) 10 Bends

Figure 2(a) to (e) shows horn attached to 2, 4, 6, 8 and 10 numbers of bends. The plain horn is of length 150 mm to which bends are attached. The mouth diameter of the horn is 50 mm and the throat diameter is 10 mm. The bends that are attached are of the same diameter as that of the throat diameter of the horn. When compressed air passes through the horn; at bends, air gets further compressed[3] resulting in greater SPL. However due to greater length of travel, SPL gets reduced. So, just before entering the flaring part of the horn the throat diameter is reduced to 7 mm. This results in further compression of air resulting in higher SPL at the outlet of the horn. Table 1 shows the overall length of the modified horn geometry considered for all the cases.

Table 1: Overall Length of Different Geometries of Acoustic Horn

No. of Bends	Overall Length of Horn (mm)
0	150
2	595
4	820
6	1045
8	1270
10	1495

4. SIMULATION STUDY

Simulation study of acoustic horn[4] is used to find out the amount of sound pressure level generated at a particular distance from the source of sound wave generation. Also frequency spectra are obtained for all cases. Simulation is carried out using COMSOL Multiphysics software version 4.3. Simulation was carried out for plain horn and for horn attached to 2, 4, 6, 8 and 10 numbers of bends.

Simulation of an acoustic horn requires the completion of following steps:

1. Modeling horn geometry
2. Meshing the geometry
3. Assigning boundary conditions
4. Studying results.

4.1 Modeling Horn Geometry

Horn geometry in 3D is created for all cases exactly according to the dimensions. Sound generated from a point source propagates in a spherical wave pattern. Also, acoustic soot blowers are generally installed at a distance of 1 m from the target area. Hence, a spherical domain of 1 m is created in front of the outlet of horn. An additional perfectly matched layer of 0.01 m is created as an extended part of the spherical domain. It is an artificial absorbing layer for wave equations commonly used to truncate computational regions in numerical methods to simulate problems with open boundaries.

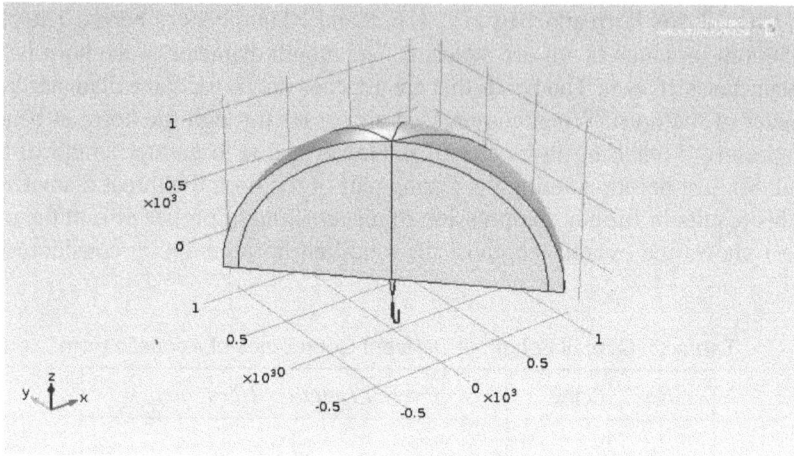

Fig. 3: Model of Horn Attached to 2 Bends

Figure 3 shows 3D model of the horn attached to 2 bends.

4.2 Meshing

Meshing is a very important step while performing simulation. In order to analyze fluid flows, flow domains are split into smaller sub-domains made up of geometric primitives like hexahedral and tetrahedral in 3D and quadrilaterals and triangles in 2D. The governing equations are then discretized and solved inside each of these sub-domains. The element selected for this simulation is free tetrahedral element. From COMSOL's mesh tool, extra fine mesh is selected.

4.3 Assigning Boundary Conditions

At the diaphragm of the horn, spherical wave radiation is assigned. When compressed air is sent through the pressurized air input of the horn, it sets the diaphragm present inside to vibration[7]. The vibration of the diaphragm results in generation of sound waves. In COMSOL we need to give the inward acceleration of the air particles due to the deflection of the diaphragm as an input.

Acceleration of the diaphragm is calculated as,

$$a = f_n^2 \times y_0 \qquad \qquad \dots (1)$$

Where, f_n = Natural frequency of the diaphragm in Hz.
　　　y_0 = Deflection of diaphragm in m.

Natural frequency of the diaphragm is calculated as,

$$f_n = \left(\frac{1}{2\pi}\right) \times \sqrt{\frac{k}{m}} \qquad \qquad \dots (2)$$

Where, k = Stiffness of the diaphragm in N/m.
　　　M = mass of the diaphragm in kg.

Stiffness of the diaphragm is calculated as,

$$k = P \times \pi \times R^2/y_0 \qquad \qquad \text{... (3)}$$

Where, P = Input air pressure in N/m^2.
   R = Radius of the diaphragm in m.

Mass of the diaphragm is calculated as,

$$m = \rho \times \pi \times R^2 \times t \qquad \qquad \text{... (4)}$$

Where, ρ = Density of the diaphragm in kg/m^3.
   t = Thickness of the diaphragm in m.

Deflection of the diaphragm is calculated as,

$$y_0 = \left(\frac{3}{16}\right) \times \frac{P(1 - \vartheta^2)}{E \times t^3} \times R^4 \qquad \qquad \text{.... (5)}$$

Where, E = Young's Modulus of diaphragm in N/m^2
   ϑ = Poisson's Ratio of the diaphragm

The values of ϑ, E, ρ, t and R for the diaphragm are given by the manufacturer of horn. Table 2 shows the value of inward acceleration of air particles at different compressed air pressures considered.

Table 2: Inward Acceleration Values of Air Particles at Different Pressures

Input Air Pressure (bar)	Inward Acceleration (m/s2)
1	4.61e3
1.5	6.91e3
2	9.21e3

4.4 Studying Results

Simulation results of horn attached to 2 bends at 1 bar

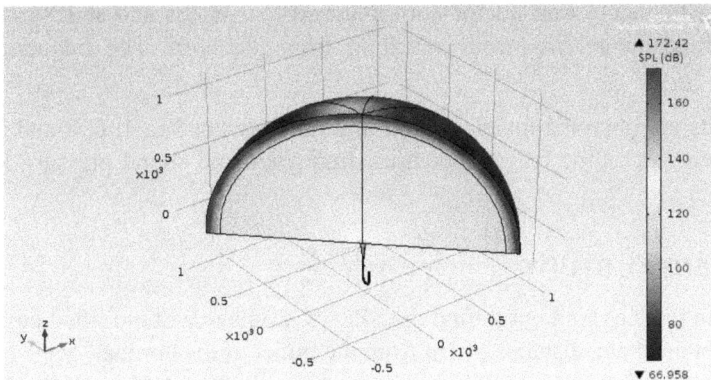

Fig. 4: Distribution of SPL over the Domain for Horn Attached to 2 Bends

Figure 4 shows the distribution of sound pressure level over the entire domain for horn attached to 2 bends and compressed air pressure is 1 bar.

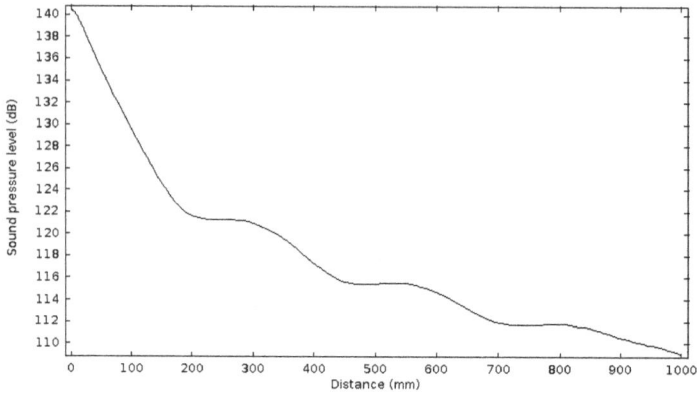

Fig. 5: SPL vs. Variation of SPL with Distance for Horn Attached to 2 Bends

Figure 5 shows that at a distance of 1 m from mouth of the horn, SPL obtained is 108.76 dB.

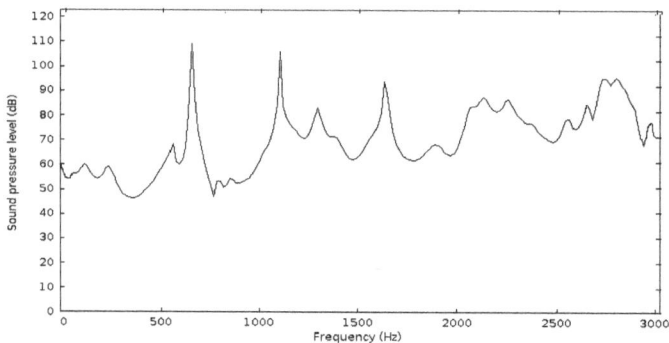

Fig. 6: Frequency Spectra for Horn Attached to 2 Bends

Figure 6 shows frequency spectra for horn attached to 2 bends and at 1 bar. The first peak indicates the fundamental frequency resulting from the horn. The frequency obtained is 729.17 Hz.

Similarly results are obtained for plain horn and horn attached to 4, 6, 8 and 10 numbers of bends. It is seen that as the length increases, frequency and sound pressure level gradually reduces.

5. EXPERIMENTAL STUDY

Experiments are to be carried out to find the SPL vs. frequency of modified design of acoustic horn. SPL is obtained at a distance of 1m from the outlet of the horn, as, acoustic soot blowers are generally mounted at the mentioned distance from the target area which is to be cleaned. Experiments are carried out in open environment.

5.1 Experimental Set-up

Experimental set-up consists of following parts:

1. Single stage reciprocating air compressor.
2. A bourdon pressure gauge.
3. Acoustic horn of desired geometries.
4. A transducer – Microphone.
5. FFT analyzer.

Figure 7 shows the line diagram of the experimental set up. The compressor used is a single stage air cooled reciprocating compressor. Bourdon type pressure gauge is used. Six cases of horn geometry are considered *i.e.* plain horn and horn attached to 2, 4 6, 8 and 10 numbers of bend. Microphone[12] acts as a transducer which converts sound waves into an analog signal. Microphone used is of Bruel–Kjaer make having frequency range of 20 Hz–20 kHz and a dynamic range of 16.5 dB–134 dB. Acoustical signal like sound waves are continuous *i.e.* they have a defined value for every possible instant of time. The sound pressure to be analyzed requires converting the signal into a stream of digital samples, with each sample representing a numeric value *i.e.* proportional to the measured signal at a specific instant of time. This process of converting an analog signal into a digital signal is called as sampling which is done using FFT analyzer.[27] The FFT (Fast Fourier Transform) is an algorithm which resolves a time waveform into its sinusoidal components. The FFT takes a block of time–domain data and returns the frequency spectrum of the data. Thus, the FFT does not yield a continuous spectrum, but returns a discrete spectrum in which frequency content of waveform is resolved into a finite number of frequency lines. The sampled time waveform input to FFT determines the computed spectrum. If a signal is sampled at a rate equal to 'f_s' over an acquisition time 'T', 'N' number of samples are acquired.

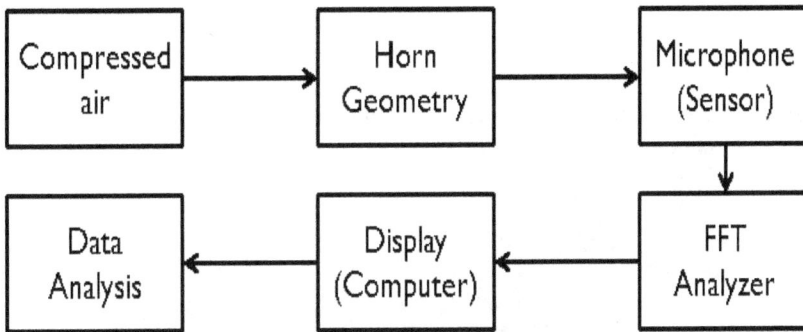

Compressed air	Horn Geometry	Microphone (Sensor)
Data Analysis	Display (Computer)	FFT Analyzer

Fig. 7: Line Diagram of Experimental Set-up

For FFT spectrum computed from the sampled signal has a frequency resolution dF, which is given by,

$$dF = f_s/N \qquad \qquad ...(6)$$

For carrying out experiments, value of f_s chosen is 6400 Hz and N considered is 400. Hence, $dF = 16$.

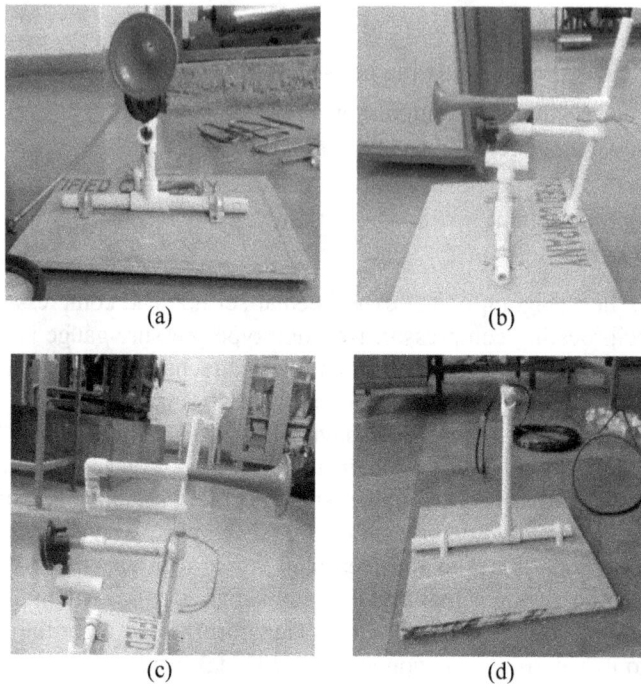

Fig. 8: (a) Horn Mounted on a Stamd (b) Horn Attcahed to 2 Bends
(c) Horn Attached to 4 Bends (d) Microphone Mounted on a Stand

Figure 8 (a) shows horn which is mounted on a stand. Outlet of the horn is seen, through which sound waves are emitted. Figure 8 (b) and (c) shows horn attached to 2 and 4 numbers of bends respectively. Figure 8(d) shows microphone mounted on a stand.

5.2 Experimental Procedure

Figure 9 shows the photograph of the experimental set-up to obtain SPL and frequency for horn attached to 2 bends. Compressed air is sent through inlet of the horn at 1 bar, 1.5 bar and

Fig. 9: Experimental Set-up

2 bar. Microphone is placed at a distance of 1 m from the outlet of the horn. Microphone is connected to the dynamic signal analyzer, at channel a1, which in turn is connected to the laptop consisting of RT photon plus software. Measurement settings are made in the software which constitutes of setting all units as per SI system, setting of dF value, choosing channel a1, selecting time capture, FFT, power spectra, octave spectra and RMS. Frequency spectra is obtained for all cases.

5.3 Experimental Results

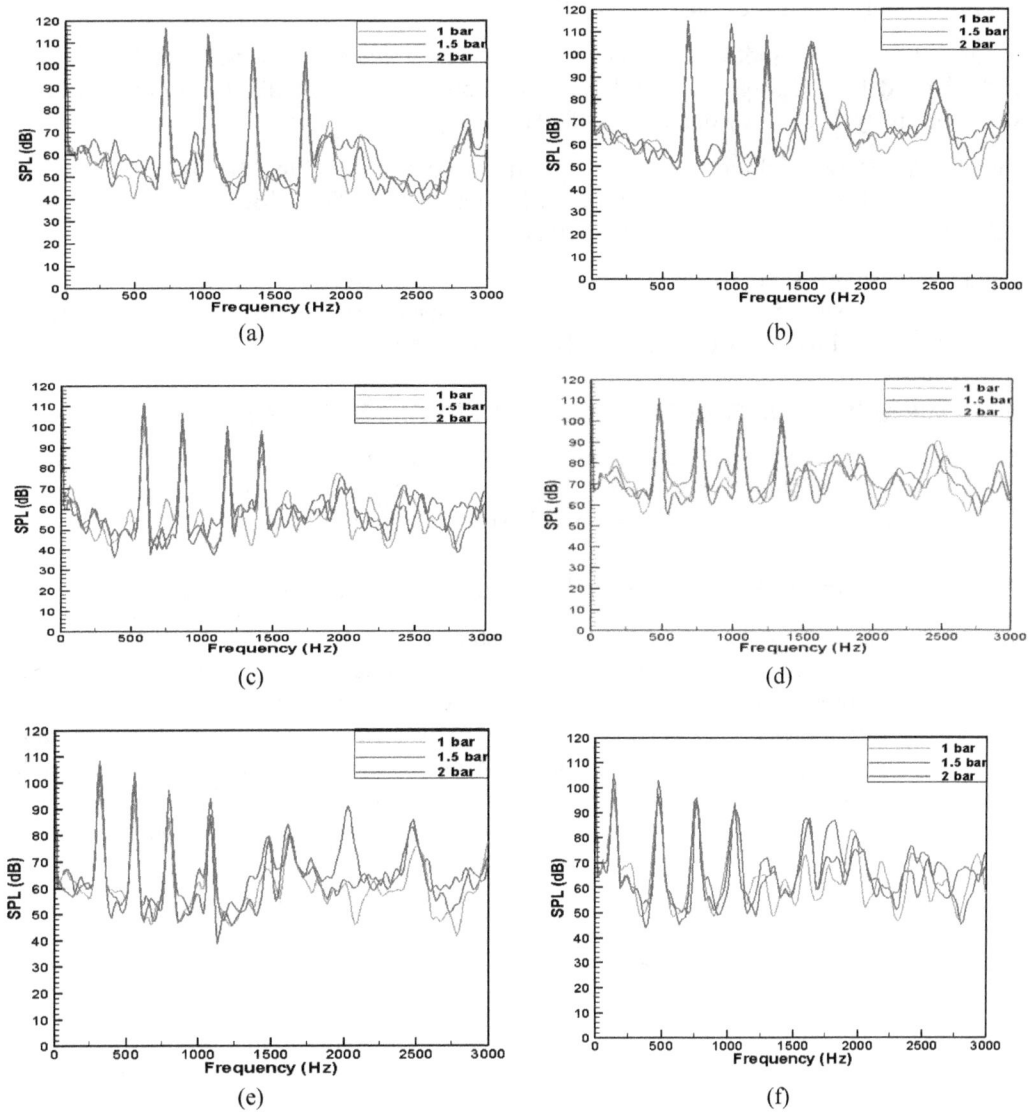

Fig. 10: Frequency Spectra for Horn Attached to (a) 0 (b) 2 (c) 4 (d) 6 (e) 8 (f) 10 Bends at Different Pressures

Figure 10(a) shows a plot of frequency spectra obtained for plain horn at 1 bar, 1.5 bar and 2 bar. At 1 bar, 1.5 bar and 2 bar, peak SPL obtained is 111.48 dB, 114.87 dB and 117.42 dB respectively. Frequency obtained is 720 Hz at all pressures.

Figure 10(b) shows a plot of frequency spectra for horn attached to 2 bends at 1 bar, 1.5 bar and 2 bar. At 1 bar, 1.5 bar and 2 bar, peak SPL obtained is 110.17 dB, 112.80 dB and 115.85 dB respectively. Frequency obtained is 688 Hz at all pressures.

Figure 10(c) shows a plot of frequency spectra for horn attached to 4 bends at 1 bar, 1.5 bar and 2 bar. At 1 bar, 1.5 bar and 2 bar, peak SPL obtained is 107.34 dB, 109.68 dB and 112.38 dB respectively. Frequency obtained is 592 Hz at all pressures.

Figure 10(d) shows a plot of frequency spectra for horn attached to 6 bends at 1 bar, 1.5 bar and 2 bar. At 1 bar, 1.5 bar and 2 bar, peak SPL obtained is 105.68 dB, 107.89 dB and 110.39 dB respectively. Frequency obtained is 480 Hz at all pressures.

Figure 10(e) shows a plot of frequency spectra for horn attached to 8 bends at 1 bar, 1.5 bar and 2 bar. At 1 bar, 1.5 bar and 2 bar, peak SPL obtained is 103.46 dB, 106.12 dB and 108.27 dB respectively. Frequency obtained is 320 Hz at all pressures.

Figure 10(f) shows a plot of frequency spectra for horn attached to 10 bends at 1 bar, 1.5 bar and 2 bar. At 1 bar, 1.5 bar and 2 bar, peak SPL obtained is 100.33 dB, 103.69 dB and 105.60 dB respectively. Frequency obtained is 144 Hz at all pressures.

6. COMPARISON OF SIMULATION AND EXPERIMENTAL RESULTS

Simulation and experimental results are obtained. The values of peak SPL and frequency obtained from simulation and experimental study are compared. Also, percentage reduction in frequency obtained from both simulation and experimentation is calculated.

6.1 Comparison of SPL Obtained from Simulation and Experimentation

Figure 11 shows SPL obtained at different pressures for horn attached to 0, 2, 4, 6, 8 and 10 number of bends obtained from simulation and experimentation. It shows variation of SPL with input compressed air pressure for all cases of horn geometry. It is seen that with increase in pressure, SPL increases. Also, percentage error in SPL is obtained by comparing simulation and experimental results. SPL obtained in experimentation is slightly more as compared to that obtained from simulation. The probable reason could be that while performing experiments background noise may add up to the actual SPL. Percentage error obtained is within acceptable limits.

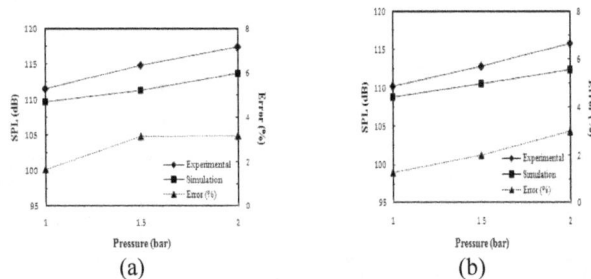

(a)   (b)

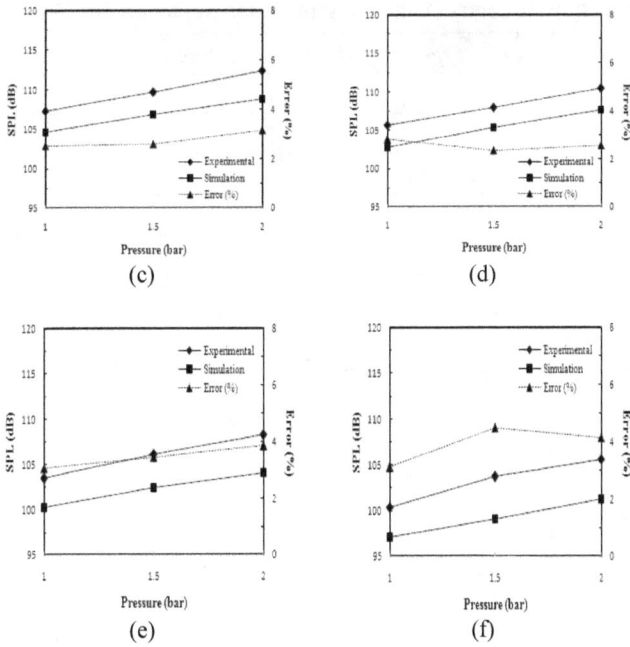

Fig. 11: SPL Obtained at Different Pressures for Horn Attached to (a) 0 bend, (b) 2 bends, (c) 4 bends, (d) 6 bends, (e) 8 bends and (f) 10 bends

6.2 Comparison of Frequency Obtained from Simulation and Experimentation

Figure 12 shows frequency at different lengths obtained from simulation and experimental results. It is seen that, frequency reduces as overall length of acoustic horn increases.

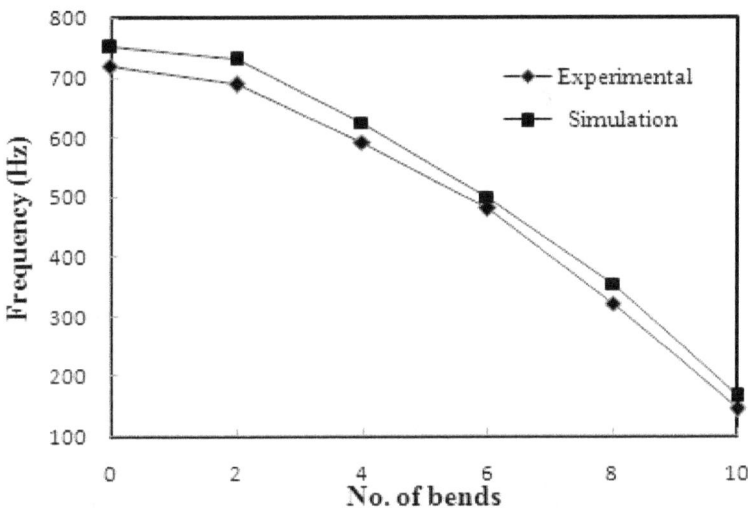

Fig. 12: Frequency Obtained at Different Lengths

Table 3: Frequency Obtained from Experimentation and Simulation

No. of Bends	Experimental Frequency (Hz)	Simulation Frequency (Hz)	% Error
0	720	750	−4.17
2	688	729.17	−5.98
4	592	625	−5.57
6	480	500	−4.17
8	320	354.17	−10.68
10	144	166.17	−15.40

6.3 Comparison of SPL with Increase in Length and at Different Pressures

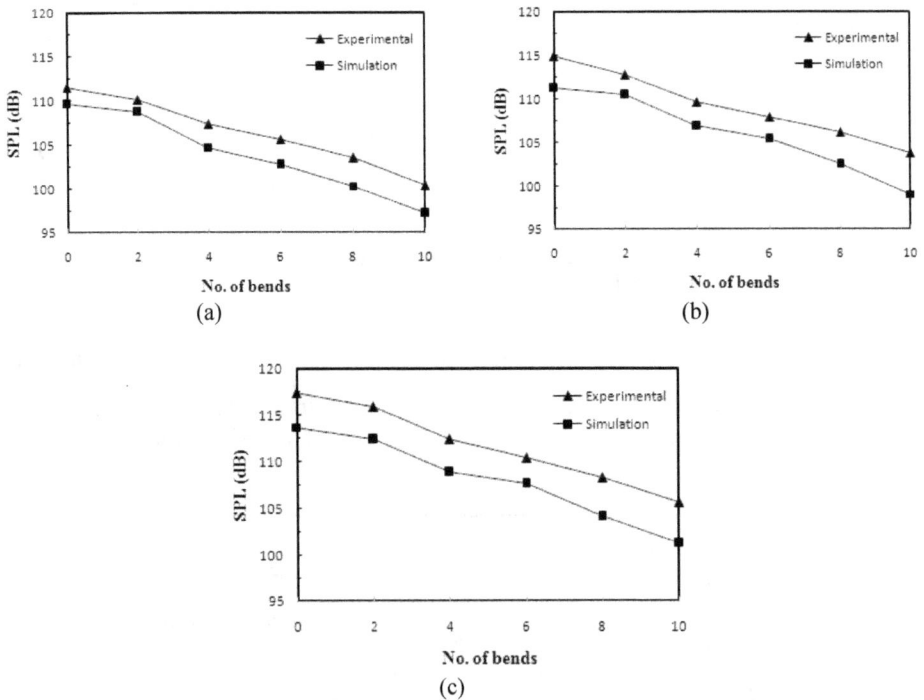

Fig. 13: SPL at Different Lengths at (a) 1 bar (b) 1.5 Bar (c) 2 Bar

Figure 13 (a), (b) and (c) shows variation of SPL at different lengths and at different pressures obtained from experimental and simulation study. It is seen that as length is increasing SPL is gradually reducing. As input air pressure increases, SPL increases.

6.4 Percentage Reduction in Frequency Obtained with Increase in Length

Figure 14 (a) and (b) shows percentage reduction in frequency vs. increase in the number of bends when compared with the frequency of plain horn. As length of the acoustic horn is increasing, frequency is reducing.

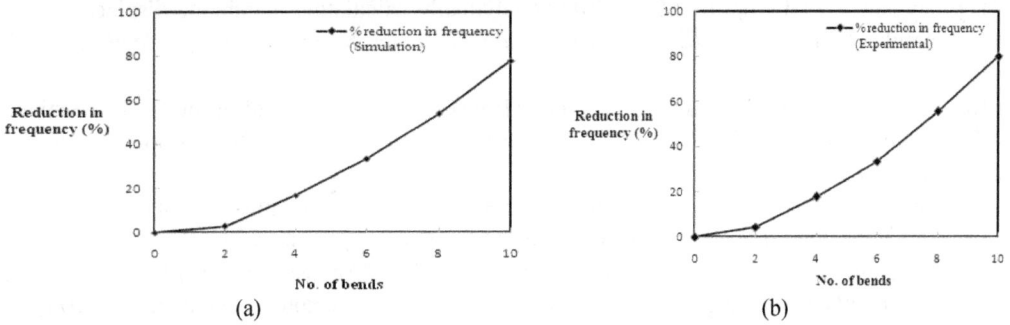

(a) (b)

Fig. 14: Percentage Reduction in Frequency (a) Simulation (b) Experimental

7. CONCLUSIONS

It is seen that, when the overall length of the acoustic horn is increased, the SPL and frequency both decreased. Similarly, when the pressure is increased, the SPL also increased. It is further observed that frequency is independent of pressure for the same length. The maximum percentage reduction obtained in frequency is about 80% for horn attached to 10 bends when compared to plain horn. On comparing simulation and experimental results, it is found that they are in good agreement and within acceptable limits.

The frequency obtained in the cases of horn attached to 6, 8 and 10 number of bends *i.e.* 480 Hz, 320 Hz and 144 Hz respectively fall within the range of frequency of acoustic soot blowers. The maximum SPL obtained is 110.39 dB which is less than the desired SPL for effective cleaning. This is because, the maximum compressed input air pressure used is 2 bar in order to protect our ears. However, in plants compressed air sent is upto 20 bar. At this high pressure, desired SPL will be obtained. The modification done in the geometry of acoustic horn successfully helped to achieve the aim

REFERENCES

[1] Naganuma, H. (2013). "Reduction mechanisms of ash deposition in coal and/or biomass combustion boilers", *Journal of fuel processing technology,* Vol. 106, pp. 303–309.

[2] Mirek, P. (2013). "Field testing of acoustic cleaning system working in 670 MW CFB boiler", *Journal of chemical and process engineering*, Vol. 34, No. 2, pp. 283–291.

[3] Dequand, S. (2003). "Acoustics of 90 degree sharp bends", *Journal of sound and vibration*, Vol. 89, pp. 1025–1037.

[4] Bangtsson, E. and Noreland, D. (2002). "Shape optimization of an acoustic horn", *Comsol Multiphysics tutorial, acoustics module,* pp. 2002–2019.

[5] Seiffert, G. and Barry Gibbs (2006). "Removal of Charged Powder Deposits by High Intensity Low Frequency Sound: The Role of Inertial and Drag Forces", *13th International Conference on Sound and Vibration,* Vienna, Austria, pp. 2–6.

[6] Yu, G., JIN, B.S. and XIAO, G. (2009). "The Fouling Characteristics and Comparative Analysis of Cleaning Technology of SCR", *11th International Conference on Electrostatic Precipitation,* Hangzhou, pp. 624–626.

[7] Wygant, I. and Kupnik, M. (2008). "Analytically calculating membrane displacement of a circular CMUT cell", *IEEE International Ultrasonics Symposium Proceedings,* pp. 2111–2114.

[8] Buhl, L., Steen Drue and Leif Lind, (1996). "Field Testing of Acoustical Cleaning of Electrostatic Precipitators", *ICESP VI Conference on Air Toxics Series*, Budapest, Hungary, pp. 18–21.

[9] Jing, T. (2004). "Applications of the Sonic Soot Cleaning Technique in Boilers", International conference on ultrasonics, ferrorelectrics conference, Beijing, pp. 2211–2212.

[10] Ronghui, W., Han Ke and Tan Ruitian, (2004) "Acoustic Horn Made Electrostatic Precipitator Collecting Plate and Hopper Clean in Zhanjiakou Power Plant", *International Conference on Electrostatic Precipitation, Lianoning zhongxin Automatic Instruments Co. Ltd.,* China, pp. 1–6.

[11] Alton, F. and Pohlmann, K. "Master handbook of acoustics", Fifth edition, Mc Graw Hill, pp. 3–38.

[12] Beranek, L. (2012). "Acoustics: Sound fields and transducers", Chaoter 9, Academic press, San Deigo, pp. 407–448.

[13] King, M. (2008). "Horn physics", Section 5, pp. 1–29.

[14] Hatt, R. (1990). "Understanding Boiler Slag", *Coal combustion. Inc*, Versailles, p. 3.

[15] Krasil'nikov, V.A. (1963). "Sound and ultrasound waves in air, water and solid bodies", pp. 26–27.

[16] Hall, R. (1984). "Sonic cleaning device and method", United States Patent Publication, Patent No. 44, 61, 651.

[17] Zhang, T. (2012). "Acoustic cleaning device with variable length to compensate application temperature", *United States Patent Publication,* Patent No. 1, 45, 182.

Increasing Corporate-Community Consistency by Entrenching Micro-Plan in CSR Policies with Reference to OPGC Limited

Ashwini Kumar Patra[1] and Manoj K. Dash[2]

[1]Department of Human Resource Management and Organization Behaviour, Rourkela Institute of Management Studies (RIMS), Institutional Area, Gopabandhu Nagar, Chhend, Rourkela–769015, Odisha

[2]Department of CSR and Sustainability, Odisha Power Generation Corporation (OPGC) Ltd., Chandrasekharpur, Bhubaneswar–751023, Odisha

E-mail: [1]patraashwini@gmail.com; [1]ashwinipatra@yahoo.com; [2]manoj.dash@aes.com; [2]sustainability@opgc.co.in

ABSTRACT: *Success of any human development programme to a significant extent is determined by judicious crafting of programme components that aim at addressing local problems and people's needs. The said process enhances people's participation, acceptability and perspectives in favour of the organization that works on such a framework. An organization that sincerely makes efforts towards creation of general well-being for its stakeholder communities through its Corporate Social Responsibility (CSR) interventions consist of conscientious actions that integrate economic, social and ecological concerns in their business operations and in their interactions with the community members. In order to convert these actions into productive, verifiable and sustainable results, an organization can use micro-plan as a tool and incorporate it in its CSR Policy towards ensuring an effective corporate-community interface. This paper draws some lessons from the means through which CSR activities are planned in the organization under study, identifies vulnerable areas and suggests suitable strategic interventions by the organization for improvements in socio-economic status of the community members. The suggested strategic interventions, as it is stated in conclusion, would lead to strengthening of mutual trust between community and the corporate organization; community empowerment and programme sustainability while contributing towards the emergence of a futuristic approach based on strong community-corporate bonding.*

Keywords: Acceptability, Framework, CSR, Programme Activities, Strategic Interventions, Sustainability, Corporate-Community Bonding.

1. INTRODUCTION

In this age of global competition, corporates are beginning to realize the stake that it has as a part of the society. There is a growing realization that they should contribute to social activities globally with a desire to improve the immediate environment, where they work and many organizations are taking keen interest in such activities (Shinde, 2005). It is also observed that to a growing degree companies that pay genuine attention to the principles of socially responsible behaviours are also favoured by the public and preferred for their goods and services. This has given rise to the concept of Corporate Social Responsibility (CSR).

Corporate Social Responsibility (CSR), also known as corporate responsibility, corporate citizenship, responsible business, Sustainable Responsible Business (SRB), or corporate social performance, is a form of corporate self-regulation integrated into a business model. Ideally, CSR policy would function as a built-in, self-regulating mechanism whereby business would monitor and ensure its support to law, ethical standards, and international norms. Consequently, business would embrace responsibility for the impact of its activities on the environment, consumers, employees, communities, stakeholders and all other members of the public sphere. Furthermore, CSR-focused businesses would proactively promote public interest by encouraging community growth and development, and voluntarily eliminating practices that harm the public sphere, regardless of legality. Essentially, CSR is the deliberate inclusion of public interest into corporate decision-making, and the honoring of triple bottom line: people, planet, profit (Kaur, 2012).

The term "CSR" came in to common use in the early 1970s, after many multinational corporations formed, although it was seldom abbreviated. The term stakeholder, meaning those on whom an organization's activities have an impact or who could impact the organization, was used to describe corporate owners beyond shareholders as a result of an influential book entitled *Strategic Management: A Stakeholder Approach* by R. Edward Freeman in 1984 (Boston: Pitman).

ISO 26000 is the recognized international standard for CSR (currently a Draft International Standard). Public sector organizations (the United Nations for example) adhere to the triple bottom line (TBL). It is widely accepted that CSR adheres to similar principles but with no formal act of legislation. The UN has developed the Principles for Responsible Investment as guidelines for investing entities.

Based on these standards, guidelines and practices, the term Corporate Social Responsibility (CSR) is defined as "A responsible business is achieving commercial success in ways that honour ethical values and respect people, communities and the natural environment. These businesses minimize any negative environmental and social impacts and maximize the positive ones."

2. SIGNIFICANCE OF MICRO PLAN AND ITS IMPORTANCE IN DEVELOPMENTAL ACTIVITIES

Micro planning is a grassroots level process, where all stakeholders but specifically the disadvantaged persons of the society get an opportunity to express their opinion and participate in their own development. The community as a stakeholder of the development process, can find their participation embedded in planning, monitoring to execution of the process. This total process helps the community to understand, express their concerns and integrate their responsibility towards sustainability of the programme.

In this process, the community members identify the vulnerable areas, design the programme activities, and involve themselves in materializing the programme. They own and lead the programme to success. In a nutshell, the micro planning process is defined as *"A decentralized process, where people come together and rechristen their own way of empowerment in the society".*

At the beginning of introduction of the process, the implementing agency consults with the community members, take note of their concerns, design and implement the programme along with the members. As the community is given priority from beginning of the programme, hence the members own the result of the program and this pattern is followed year after year, generation after generation, unless and until any impeding situation compels them to bring about changes in the planning, implementation and monitoring process.

3. NEED OF INTEGRATING MICRO PLAN IN CORPORATE SOCIAL RESPONSIBILITY (CSR) ACTIVITIES

It is expected that in all types of development interventions micro planning process should be followed and people's participation is an important component of the said process. As stated in Section 135 of the Companies Act, 2013, it is mandatory that all Corporate Organizations fulfilling specific conditions must shoulder the responsibility towards holistic development of their stakeholder communities. Hence, in order to get positive outcomes from the amount invested/spent in corporate social responsibility activities, it is imperative for the corporate organization to incorporate micro plan as a process in its Corporate Social Responsibility (CSR) Policy.

While deriving CSR policy of a corporate organization, it is the concern of top level management or Board of Directors to maintain a balancing act among the three components, *i.e.*, people, planet and profit. People stand for social well-being; Planet for ecological quality; and Profit for economic prosperity. Balancing of the three P's in the organization means creating an opportunity for the community to make choices and set their priorities in designing the programmes/activities as a part of its CSR plan/strategy. Thus, such a policy provides a view of the thought process of the corporation and the role it assumes in the society. It represents the core responsibility of firms to pursue goals in addition to profit maximization and the responsibility of a firm's stakeholders to hold the firm accountable for its actions. Further, the process embodies corporate, economic, social and environmental indices, carried out for the benefit of the stakeholders and the society, and not for the mere achievement of economic growth, return on investment or minimum compliance with the legal framework.

The aim in 'micro-planning' is on planning from the lowest level *i.e.* from the functional community upward to a clearly defined region to fulfill the need of the local areas and ensuring the process of integration of the different areas with an objective to attain balanced regional development. Therefore, location of specific socio-economic activities and their inter-linkage over a region or particular geographical area are the major concerns of micro-level planning.

Micro-planning takes into cognizance the evolution of the spatial pattern of human activities without which economic, social and environmental goals of planning cannot be achieved upto expectation. It is thus put greater emphasis on those sectors which support the people of lower income groups, particularly the poor and the weaker sections in rural areas with an aim to offer them a better quality of life and ameliorating their deplorable socio-economic conditions (Shah, 2013).

In this context, the following theoretical model describes the benefit of incorporating micro-plan while designing the CSR Policy of a corporate organization.

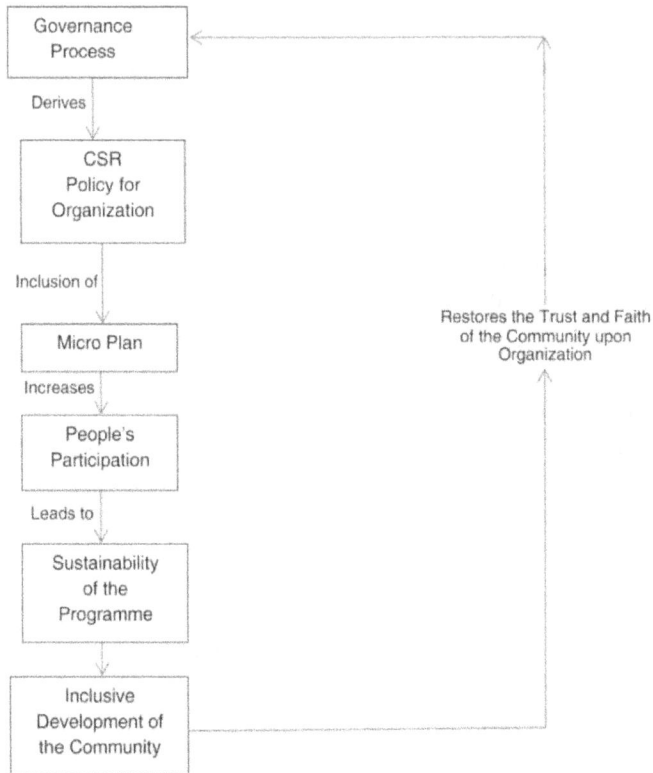

Fig. 1: Theoretical Model of Micro Plan
(Model Developed by the Authors)

In this paper, the focus is to study the following aspects:

- Find out the level of acceptability of programme activities among the community members and their participation in programme planning, monitoring and implementation.
- Ascertain the role of community in deriving a sustainable strategic plan of various thematic activities initiated by OPGC.
- Find out magnitude of community participation in the CSR Policy formulated by OPGC.
- Suggest and recommend improvement in CSR Policy for better result in community participation, mutual trust improvement and tangible impact on society.

4. HIGHLIGHTS OF CSR POLICY OF ODISHA POWER GENERATION CORPORATION

The CSR Policy of OPGC was framed by its CSR Committee in the year 2014 which was then approved by its Board of Directors as per provisions of Section 135 of the Companies Act, 2013, and CSR Rules of 2014. The highlights are as follows:

Objectives

- To uphold and promote the principles of inclusive growth and equitable development for stakeholder communities based on needs and priorities;
- To contribute as well as facilitate holistic development of stakeholder communities through participatory planning and accountability as well as transparent processes by measuring effectiveness of our programmes. The thrust would be on gradually moving away from philanthropy towards sustainability;
- To work actively in the areas of health, education, nutrition, drinking water, sanitation, vocational skills for employability, livelihoods and income generation for empowerment of women and youth, creation and development of community infrastructure (e.g. roads, educational facilities, etc.) for rural development, water resource management and water conservation and training of children/youth in sports. All these projects will be undertaken with an aim to enhance human development index of stakeholder communities around OPGC's operational areas defined and decided by its Board from time to time;
- To promote clean environment in the communities near its operational areas;
- To encourage volunteering among internal and external stakeholders for robust bonding with local communities and other stakeholders;
- To collaborate/partner with professional bodies like government organisations and reputed NGOs having well established track record of at least three years and established academic institutions towards leveraging resources, implementation of CSR programmes and pursuit of other objectives;
- To interact regularly with its stakeholders, review, publicly report the outcomes of CSR initiatives and share the best practices.

Strategy

- Promoting good CSR governance;
- Projects aligned with CSR Rules and OPGC's CSR objectives;
- Monitoring and measurement of CSR projects.

One may see that, 'participatory planning' is recognized as one of the objectives of OPGC's CSR Policy. The country and its corporate organisations are new to a structured approach to CSR planning and interventions. OPGC is also no exception to this general prevailing condition. Hence, a structured micro-planning process in OPGC's CSR framework will take its natural course while developing into a robust and self-repeating form.

As one can see in the model presented above, the decision making process for CSR interventions of OPGC is propelled by multi-dimensional forces. Community as the primary stakeholder puts forth their priorities to be fulfilled at the village level, and these are facilitated by local elected representatives, District Administration, and local pressure groups. To supplement these, priority identification assessments are carried out at the grassroots level and compliance requirement audits are carried out by the members of OPGC's CSR team in its operational area for larger benefit of the community.

The CSR activities of OPGC focus on two broad areas: physical infrastructural development for improving quality of life and enhancing human development indices to augment well-being.

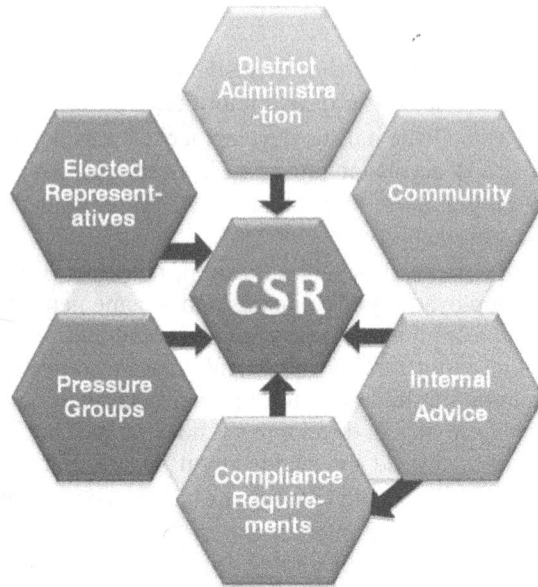

Fig. 2: Model of Determining CSR Interventions in OPGC
(Model Developed by the Authors)

In physical infrastructural development, prominent activities like developing sanitation and drinking water supply systems, pond construction or renovation, internal village roads, street lighting systems, multi-purpose community buildings in their habitats, etc. are carried out in all stakeholder villages on community's demands. While fulfilling these demands, the community members are asked to present a detailed plan of present numbers of beneficiaries and their future plan for up-keeping of these community resources.

The human index development activities of OPGC are more focused upon empowerment process of community and their contribution in sustenance of the activities. In this category, long term needs of the community are addressed. The communities realized in the course of time that issues like handling of water scarcity and lack of a sanitized environment pose a big threat to the survival of life and sustenance of livelihoods. In order to mitigate these threats, long term programmes like water and sanitation activities supplementing the current Swwach Bharat Abhiyan, skill building among youth to leverage micro entrepreneurship opportunities and augmenting per capita income of the rural households, strengthening livelihood activities to increase the socio-economic status of the family and supporting them to maintain a good quality of life, and strengthening the base of quality primary education at the school level have been initiated with prior community consultation and utilizing the capacity of the CSR team to plan different interventions to meet the needs as well as aspirations of different stakeholder groups in the community.

The strategy formulated in deriving and executing CSR policy of OPGC is of twofold: first is direct involvement in the work. OPGC is directly involved in carrying out the infrastructural activities. Second, development of credible partnerships and execution of time-bound and result oriented programmes with the professional support of well-known professional organizations in different fields of intervention, *i.e.* education, livelihoods and water/sanitation. The human development activities are executed in the field with the support of partner organizations. These strategic thrusts in the CSR Policy of OPGC gives the organization an upper hand for effective monitoring and accelerating the pro-humanistic objectives of the organization, *i.e.*, harmonious co-existence of both the industry and community; and maintaining an improved quality of life by the community.

5. METHODOLOGY

In order to achieve the objectives of this study, it is important to examine the extent of involvement of beneficiaries in planning, monitoring and execution of CSR activities initiated by OPGC in its operational area, satisfaction level of beneficiaries in regard to theme specific prioritization of problems (*e.g.,* infrastructure, education, water and sanitation, skill development, etc.) and operational procedures in addressing the problems, empowerment process and involvement of people in programme sustainability, and establishment of rapport of OPGC people with the beneficiaries living in the stakeholder habitats.

In broad terms, three types of analyses are planned in the study. These are:

- Flexibility in CSR Policy framed by OPGC where priority of the people and their participation are given emphasis.
- Whether investments made by the corporate organization under its CSR activities has major people component.
- The outcomes of investment of resources in respect to holistic benefit to the community as well as organizational benefit.

The study has made use of both quantitative and qualitative methods to explain effectiveness of CSR Policy formulated by the organization, benefits reached out to the targeted beneficiaries, ongoing activities to mitigate the chronic problems of the areas. The study has made use of a short but structured interview schedule regarding acceptability of the programme activities and its usefulness to address the recurring problems of the beneficiaries and its future prospects. Accordingly, 100 beneficiaries were randomly selected from 10 peripheral villages of the OPGC out of a total of 40 villages. The beneficiaries were asked questions on frequency of consultation of officials with the community, priority given to assessment of the problem before initiation of any activity, wide scale acceptability of the programme among the members of the community, awareness among the community members about different programmes and its benefit to the society as a whole, management of the programme by the community to fulfill its present objectives and continuation in future, and support extended from the OPGC to continue the interventions to enhance the living standards of the community as well as inclusive growth of the society. For each and every question, a five point scale was developed. The respondents were briefed about the study and translated in vernacular medium for thorough understanding of the questions. The interview schedule was administered in the identified operational villages of OPGC by trained professionals. The opinions of the corporate officials engaged in CSR activities were also collected regarding future plan of the organization, present focus of its CSR Policy and implementation process. The collected information was analyzed with the use of simple statistical tools like calculating highest percentage of beneficiaries who responded to the question on a five point scale as already mentioned.

6. MAIN FINDINGS OF THE STUDY

It is observed that fifty three percent of respondents agreed on the issue that before launching of any of the programme by OPGC, the problems of the people are taken into consideration and it is prioritized in consultation with members of the community. Based on the problem prioritization, activities are planned in consultation with the beneficiaries and implemented. Taken together with those who strongly agreed, the percentage goes up to an overwhelming seventy two percent who agreed with the statement. In contrast, thirteen percent and eight people expressed simple disagreement and strong disagreement with the issue respectively.

Table 1: Distribution of Respondents on Prioritization of Programme
in Operational Villages (in percentage)

Response	*Percentage*
Strongly Agree	19 (19)
Agree	53 (53)
Neither Agree nor Disagree	7 (7)
Disagree	13 (13)
Strongly Disagree	8 (8)
Total	**100 (100)**

Note: The number in parenthesis denotes the number of respondents.

It is observed from the above table that seventy six percent of respondents have strongly agreed on the issue that the awareness level of beneficiaries on thematic intervention areas has increased after programme intervention. If the people who simply agreed with this statement is taken into consideration, the percentage increases to ninety. They realized the benefits of their participation in the programme and accordingly involvement in the programme is increased. On the opposite, those who disagree represent only eight percent of the total.

Table 2: Distribution of Respondents towards Awareness Level and
Participation of Beneficiaries on Programme Planning and Execution (in percentage)

Response	*Percentage*
Strongly Agree	76 (76)
Agree	14 (14)
Neither Agree nor Disagree	0 (0)
Disagree	8 (8)
Strongly Disagree	2 (2)
Total	**100 (100)**

Note: The number in parenthesis denotes the number of respondents.

Table 3: Distribution of Respondents Towards Enhancing the Quality of Life
of Beneficiaries Through the Programme Contents (in percentage)

Response	Percentage
Strongly Agree	6 (6)
Agree	21 (21)
Neither Agree nor Disagree	48 (48)
Disagree	17 (17)
Strongly Disagree	8(8)
Total	**100 (100)**

Note: The number in parenthesis denotes the number of respondents.

It is observed from the above table that forty eight percent of respondents are neutral about the benefit of the programme percolating in enhancing their socio-economic status, as well as spending pattern for different needs of their households and enhancing the quality of life in society. The percentage is twenty one percent of those who realized that the activities launched by OPGC will help them in enhancing their economic status, increase their spending pattern focusing on different indices of human development and to maintain a better standard of living in the society. However, seventeen percent of respondents expressed disagreement about quality of life to get enhanced because of the interventions of OPGC under its CSR plan.

Table 4: Distribution of Respondents Towards Provision of Support Services
by the Organization for Continuation and Sustenance of Activities (in percentage)

Response	Percentage
Strongly Agree	67 (67)
Agree	23 (23)
Neither Agree nor Disagree	0 (0)
Disagree	7 (7)
Strongly Disagree	3 (3)
Total	**100 (100)**

Note: The number in parenthesis denotes the number of respondents.

It may be observed that sixty seven percent of respondents have strongly agreed that OPGC is providing support services to carry out both physical infrastructural and activities responsible for developing human indices. The organization has built capacity of the members of the community at the grassroots level to continue the interventions or maintain the assets by forming community organizations. The process empowers them for smooth management of the activities and setting up procedures for continuation of same activities in future without facing any major hurdles. The contrasting value is very minimal, as only seven percent of people have expressed their opinion to the contrary.

Table 5: Distribution of Respondents Towards Participatory Monitoring of Activities
for Deriving Better Results (in percentage)

Response	*Percentage*
Strongly Agree	9 (9)
Agree	36 (36)
Neither Agree nor Disagree	27 (27)
Disagree	19 (19)
Strongly Disagree	9 (9)
Total	**100 (100)**

Note: The number in parenthesis denotes the number of respondents.

One may observe that thirty six percent respondents have agreed on the issue of presence of participatory monitoring of activities which have been initiated in their villages after advocacy by OPGC to create such a mechanism. They have opined that the process of combined monitoring plays a pivotal role in better implementation of activities, in chalking out the role and contribution of community members in implementation of the activities, meeting the target and achieving the desired objectives of the project. In comparison to that, only twenty seven percent of respondents have remained neutral on this issue and nineteen percent of the respondents have expressed disagreement on existence of this process on ground.

Table 6: Distribution of Respondents Towards Opinion Regarding Role of OPGC and Community
in Phasing Out and Sustainability of the Activities under CSR Policy (in percentage)

Response	*Percentage*
Community Empowerment	12 (12)
Creation of Community Fund	28 (28)
Continuation of OPGC in the Villages Forever	28 (28)
Formation of Apex Level Agency	9 (9)
Ascertain of Livelihood Source	23 (23)
Total	**100 (100)**

Note: The number in parenthesis denotes the number of respondents.

It is observed from the above table that twenty eight percent of the respondents have opined regarding creation of community fund in the villages, which will enable them to continue the programme for ever, whereas the same number of respondents have opined about continuation of welfare activities by OPGC forever. At the same time, twenty three percent of respondents gave their opinion regarding ascertaining livelihood source for the members of the community through various skill building programmes which will assure them to maintain an improved life standard and continuation of the activity in future. Twelve percent of respondents emphasized upon empowering the community members through which they will able to tap the resources from various sources and carry out the activities in future.

7. CONCLUSION

The micro-planning and participatory approach outlined in the CSR Policy of OPGC highlighted in various ways throughout this paper is aimed at thorough understanding of the community's problems and challenges by the corporate organization, involvement of community as a primary stakeholder in addressing the problem, deriving a mechanism to empower them and streamlining it for sustainability of the programme.

The governing body of the organization (Board of Directors) must emphasize that human development component remains at the top while preparing various interventions as per its CSR policy. This component must have the characteristics to reflect upon a process oriented approach that integrates micro-planning throughout. Based on it, steps are to be taken to make the people aware on various issues people are empowered to make their own decisions regarding development interventions and ensure a significant level of involvement in every activity. The process will help the community to improve their financial status and recognition in the larger society. Moreover, this process will contribute towards increasing the community towards the business organization and the business organization would be able to win the trust of the community. Hence, there is simultaneous growth of the community as well as the business organization by operating together and through better bonding.

From analyses of the above data, it is observed that community members have shown positivity towards creation of community institutions and mechanisms to ensure continuity of the activities and assets built for their well-being. The community members have emphasized strengthening of their sources of livelihood for improvement of their standard of living. The top management of OPGC must emphasize more on livelihood programmes for the community members, which must be planned along with the members. Further, the combined monitoring yields better results, so it must be continued or further strengthened. The organization must upscale its interventions towards empowering the community, which will create sustainability for the organization's business. The people have a good level of satisfaction as the organization is assessing the areas of vulnerability and initiating activities which are prioritized by the community to address the identified vulnerabilities. It is also evident that the organization is providing support services which needs augmentation andthe current mechanism must be strengthened to ensure the community's contribution in each thematic area of intervention for sustenance of all activities. This will ensure that the stakeholders will be adequately empowered to take their own decisions for their development.

8. LIMITATION

The study explains the CSR planning, monitoring and execution process in one organization, *i.e.* OPGC; instead it would be better to compare the process followed in another organization of similar nature and mapping the opinions of people on all identified common issues related to both organizations.

REFERENCES

Carroll, B. Archie (1979). "A Three-Dimensional Conceptual Model of Corporate Performance," *The Academy of Management Review*, Vol. 4, No. 4 (October), pp. 497–505.

Devarani, Loukham and Debabrata Basu, Corporate Social Responsibility–Some Basic Dimensions (https://www.google.co.in/?gfe_rd=cr&ei=J1jQVc2iJKLv8weQmpfwCw#q=Devarani%2C+Loukham+and+Debabrata+Basu)

Friedman, Milton (1970). "The Social Responsibility of Business is to Increase its Profits," New York Times Magazine, September 13.

Government of India, The Companies Act, 2013 (http://www.mca.gov.in/Ministry/pdf/Companies Act2013.pdf).

Government of India, Companies (Corporate Social Responsibility) Rules, 2014 (http://www.mca. gov.in/Ministry/ pdf/CompaniesActNotification2_2014.pdf).

Kaur, Vikramjit (2012). Corporate Social Responsibility: Overview of Indian Corporates, *International Journal of Management and Social Sciences Research*, (IJMSSR), ISSN 2319-4421, Vol. 1, No. 3 (December), pp. 48–54.

Odisha Power Generation Corporation, CSR Policy, 2014. http://www.opgc.co.in/plc/csr-policy.asp

Shelly, Shah (2013). Micro-Planning: Needs, Aims and Objectives of Micro-Planning. http://www.sociologydiscussion.com/planning/micro-planning-needs-aim-and-objectives-of-micro-planning/1085.

Shinde, S. (2005). Social Responsibility Corporate style. (http://www.expresscomputeronline.com/20050502/technologylike01.shtml).

Wood, D.J. (1991). Corporate Social Performance Revisited, Academy of Management Review, Vol. 16, pp. 691–718.

A Conceptual Framework of Mastering Art of Competitive Advantage through Good Corporate Governance

A. Lakshmana Rao

College of Management and Economic Studies, University of Petroleum
and Energy Studies, Dehradun, India
E-mail: alakshman@ddn.upes.ac.in; indialax@gmail.com

ABSTRACT: *Corporate Governance is a buzz word in the field of economic administration, regulatory framework and behavioral sciences. The subject of corporate governance has its relevance and significance to varied stakeholders in different ways. In fact, Corporate Governance is a form of obligation, which a corporate body has towards shareholders, employees, customers, Government, Public and towards the Society. Organizations, which are known for good governance by fulfilling all these obligations with a proper blend, are the lead players for the others to follow for securing better and effective competitive advantage. Keeping in mind these varied obligations, organizations and corporate bodies regularly update their policies and practices especially for continued competitive advantage but the process of updating is not an easy affair, they have to find it in a pro-active manner to withstand in the market.*

The present research paper with this in view aimed at understanding the framework of corporate governance and its role in securing better and effective competitive advantage from the ambit of various stakeholders with a broader consideration from the angle and obligation of Sustainability and Corporate Social Responsibility. Further, the study remarked the changing nature obligations for existence of corporate bodies under dynamic environment. The research paper also differentiated the gap between theory and practice in adoption of sustainability practices. Finally, the research paper ends with some suggestions and ways for better and good governance for organizational sustainability.

Keywords: Competitive Advantage, Corporate Governance, Obligations, Stakeholders, Sustainability and Corporate Social Responsibility.

1. INTRODUCTION

Organizations are managed since their inception with a mechanism. The mechanism to run this show is Corporate Governance. Though the term may not be called in today's parlance, knowingly or unknowingly the corporate entities used the word Corporate Governance since time immemorial. Corporate Governance is a term related to management of corporate bodies. Corporate bodies are managed through a mechanism or system of governance. The system or mechanism consists of board of directors, shareholders, employees and other stakeholders. The relative effectiveness or ineffectiveness of management of corporate bodies is generally rests with these functionaries.

Simply speaking governing the corporate in terms of administration or management is corporate governance. The concept or framework of corporate governance got it prominence because of growing corporate scandals, misconducts, shareholder activism, Govt. control, changing role of law, ethics, and corporate social responsibilities and not but the least changing employee expectations.

Organizations are not operate in a vacuum, they are dynamic and the dynamics of nature or environment plays a vital role on it, be it in the form of internal or external environment. Corporate Governance is surrounded by a host of factors. An organization in order to gain competitive advantage or drive should consider all the factors. Some of the factors, which have prominence with respect to corporate governance (See Figure 1) are as follows:

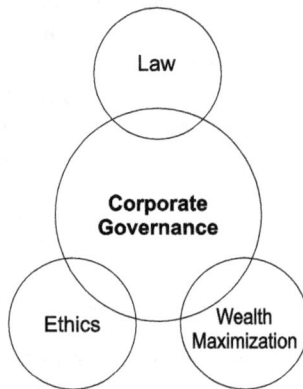

Fig. 1: The Nature of Corporate Governance

The above figure portrays the watch dog nature on the part of corporate governance, as it is aptly true that the chief aim of any organization is maximizing its wealth. The organization which lacks an eye on its governance will certainly minimize its wealth in due course of time. An organization with a frame of work corporate governance consisting of law, ethics and maximizing its wealth will definitely find a better place in terms of market growth and share by leading to a better competitive advantage as evidenced in certain research works.

In order to understand better the nature of corporate governance, it worthwhile to have broader look on the connected elements to corporate governance.

1.1 Law

Simply speaking Law Prescribes a general code of conducts to all its constituents with an obligation to follow and enforces justice with the help of certain rules and regulations prescribed by the State.

The major duty of law is enforcement of justice in terms of equity, fairness and good concise. In the case of business, it is quite common by the unscrupulous managements to deceive the innocent investors, creditors, suppliers of funds, Govt., Society and employees. Law in those cases acts as a weapon to protect their legitimate claims.

1.2 Ethics

The word 'ethics' owes its origin to Greek word 'ethos' which means a system of moral principles, the basis for deciding right and wrong actions, and implies customs, habits, noble traits in character, discipline, culture and disposition. The ethical principles are the rules of conduct that are derived from ethical values, known as six pillars *viz.* trustworthiness, respect, responsibility, fairness, caring and citizenship. These values are inseparable from one another and are closely interwoven and interlinked with each other.

It is now established that a high sense of professional morality must comprise one of the core values of corporate governance for long term as also short-term success of a company. The ethical values now-a-days, are not looked upon any longer, as costs imposed on the industry or a check on efficiency and profit maximization, but are regarded as imperatives for sustainable corporate growth and competitive edge. A highly developed sense of ethics enables corporate governance to build a trusting, long term relationship with consumers.

Ethics call for moral obligations on the part of various parties; accordingly organizations have lot of moral obligations towards various parties. Some of the moral obligations are as follows:

1. Obligations towards share holders
2. Obligations towards creditors
3. Obligations towards suppliers or vendors
4. Obligations towards customers
5. Obligations towards employees
6. Obligations towards Govt.
7. Obligations towards Society or Environment.

1.3 Wealth Maximization

Wealth maximization basically refers to maximizing the wealth of the shareholders in terms of organization. If the organization maximizes its wealth, it results in organizational development and growth. The major problem involved with this concept is weather organizations can maximize its wealth ignoring ethics?

Though the wealth maximization concept does not contain any of the obligations, yet an organization which maximizes its wealth with ethical and societal obligations in true spirit can be considered as the real follower of wealth maximization concept in practice.

To be precise, Corporate Governance in exclusion of these terms is useless. The current research paper in its analysis takes all these aspects in appropriately.

1.4 What is Good Corporate Governance?

The words "governance" and "good governance" are being used increasingly in governance literature. Bad governance is being recognized now as one of the root causes of corrupt practices in our society. Institutional investors and international financial institutions provide their aid and loans on the condition that reforms that ensure good governance are put in place by recipients. Good corporates are not born, but are made by the combined efforts of all stakeholders, board of directors, Government and the society at large (Fernando, 2009).

1.5 Theories of Corporate Governance

The corporate governance practices can be well understood with the help of theories of corporate governance. There are two prominent theories of corporate governance. They are

1. Agency theory
2. Stake holder theory.

According to Agency theory as corporate are invisible units they have to be managed with help of agents, who on behalf of a principal (corporate) manage e extended partnerships. The principal (business) delegates to the agent to carry all or majority of the powers, on their behalf to manage the affairs of the business. The relationship of agent and principal is observed in managing the affairs of such corporate bodies.

The Stakeholder theory is a very old theory. According to it Corporate Governance is a synthesis of economics, behavioural sciences, ethics and various stakeholders to business. Accordingly it considers various stakeholders such as shareholders primarily, employees, customers, dealers, Government and the Society at large. In fact, this theory stemmed out of the drawbacks of agency theory, yet this theory also consists of limitations.

2. PROBLEM STATEMENT

How to lead a better and effective competitive advantage through the adoption and implementation of framework of Corporate Governance?

3. RESEARCH OBJECTIVE

The prime objective of the current research is to determine the adoption of the societal, investor and employee obligations for enhanced competitive advantage.

3.1 Research Methodology

The current research paper as mainly conceptual, it has carried with the help of the majorly secondary sources of data, consisting of sustainability reports, annual reports, text books, reference material, journal articles, and both published and unpublished resources. The data so arrived is organized, analyzed and tabulated for better understanding of the theme and in that process it arrived with some unsolved questions for taking over the study towards a future concrete research by substantiating with help of primary data in the form of observation and interviews in select organizations for reliability.

3.2 Review of Literature

The present study is made due to the growing necessity of good corporate governance, growing social responsibility obligation, ethics and human intervention. The implications of ignorance of these elements to the maximum extent will lead to either losing of competitive advantage or its dilution. Accordingly, the study considered various themes and other early research in the field.

Donaldson, L. and Davis, J.H. (1919) advocated various theories of Corporate Governance. According to them the development of Corporate Governance is a global occurrence and, as such, is a complex area including legal, cultural, ownership, and other structural differences and accordingly some theories may be more appropriate and relevant to some countries than others, or more relevant at different times depending on what stage an individual country, or group of countries, is at or may have.

In an Article published by Ruth Aguilera and George YIP in Financial Times, May, 2007 on "Global strategy faces local constraints" argued that there are five major players who affect the company's decisions about global strategy. They are employees, board of directors, top management, Government etc., excluding other stakeholders such as customers, suppliers and competitors.

Fernando A.C. (2009) gave a detailed description of Corporate Governance Principles, Policies and Practices in theoretical and with case study orientation.

Vijaya Murthy (2004) undertook an analytical study of the corporate social disclosure practices of the top 16 software firms in India by analysing their annual reports using content analysis to examine the attributes reported relating to human resource, community development activities, products and services activities and environmental activities. It was revealed that the human resources category was the most frequently reported followed by community development activities and the environmental activities was the least reported. Most of the information was qualitative. Some firms had separate sections for each category while many others disclosed their social practices in the introductory pages of the annual report.

Jackson, I.A. and Nelson, J. (2004) in their books provide a comprehensive description of the global trends, competitive pressures, and changing expectations of society that are reshaping the rules for running a profitable and principled business. It also offers companies a framework for mastering the new rules of the game by realigning their business practices in ways that restore trust. Information is presented on the crisis of trust, the crisis of inequality, and the crisis of sustainability. The book presents the following seven principles that serve as a framework:

1. Harness innovation for public good
2. Put people at the center
3. Spread economic opportunity
4. Engage in new alliances
5. Be performance driven in everything
6. Practice superior governance
7. Pursue purpose beyond profit.

The seven principles can be used as a compass to help executives and managers navigate new terrain and apply the strategies and terminology most appropriate for each company. The book focuses on companies and business people who are delivering both private profits and public benefits. It profiles real companies delivering measurable performance and concrete solutions for stakeholders.

Ho and Taylor (2007) given statistics and findings on 50 largest US and Japanese companies based on GRI (The Global Reporting Initiative) reporting Guidelines stated that the extent of reporting is higher for firms with larger size, lower profitability, lower liquidity and for firms with membership in the manufacturing Industry.

Dutta (2011) discussed the necessity of considering the three parameter People, Planet and Profit to have a more comprehensive mechanism that integrates the traditional financial information along with non-financial information, which can help firm in enhance economic value addition, besides putting it on a firm financial footing.

Hubbard (2009) proposed Sustainable Balanced Score Card (SBSC) conceptual framework coupled with a single measure organizational sustainability performance Index to integrate the measure of SBSC to overcome TBL reporting.

In an article by **Irani, J.J.** *et al.* (2005) it is viewed that being ethical does not mean one cannot also be profitable. It is most important to make profits and to generate wealth because only then can one have the resources to do good for the community. The differentiator between good and bad business practices is what happens to the wealth after it has been generated.

Tyleca *et al.* (2002) noticed that firms standardized their environmental measurement systems respond to community demands for more transparency.

According to **Vanek, (1975)**, the idea of employee's participation emerged as an intellectual reaction to the evils of modern capitalism. The concept is deemed to provide a competitive edge in the world market place.

Chiplin, B. and Coyne, J. (1977) studied that in any society, the distribution of property rights affects the allocation and use of resources in specific and predictable ways. However, in democratic societies, ownership should not confer an unchallengeable right to exercise authority over other human beings. The advocates of employees' participation in business enterprises have propounded that ownership of property should both give rights as well as vest social responsibility.

4. CORPORATE GOVERNANCE AND COMPETITIVE ADVANTAGE

Competitive Advantage refers to the advantage enjoyed by an organization that perform some aspect of its work better than competitors or in one way competitors cannot duplicate. According to Michael Porter, the pioneer in the field of competitive advantage has argued that they are five forces that affect the potential profitability of an industry. They are industry rivalry, customers' bargaining power, suppliers, availability of substitutes and barriers to entry. These forces provide the basis for the development of an analytical tool called the value chain, which provides insights into how firms assess their capabilities to compete and make decisions regarding their competitive strategies. With respect to gaining competitive advantage is concerned, investors are the backbone of any corporate body and they are primarily interested in transparency, accountability and fair dealings, this can be possible only with good corporate governance mechanism that exists in corporate bodies. Therefore, in one way the starting point of competitive advantage is good Corporate Governance only.

4.1 HR Strategies for Sustainable Competitive Advantage

The key player in gaining competitive advantage is employees. In fact, organizations in today's scenario visualizing their employees as their core assets and further they are viewing at the time of hiring itself people from investment perspective. In order to realize their goals as enshrined in organization's vision and mission statements, organizations regularly formulate and update their Human Resource Management Strategies. HR strategies are essentially plans and programme to address and solve fundamental strategic issues related to human resources management. HR strategy focuses on the alignment of organization's corporate as well business unit plans.

Human Resources are vital to secure competitive advantage. However, securing competitive advantage is not an easier affair. Organizations have to concentrate on people dimensions. They have to raise a fundamental question: "What kinds of people will be needed to lead the organization in the years to come to have a sustainable competitive advantage?"

The establishment of an identified set of corporate values that has people's development at its epicenter is vital to the sustained development of an organization that depends fundamentally on the ideas, commitment, and motivating its human resources.

Increased profit and revenue no longer remain the key motivators of corporate success. The level of employee motivation, customer engagement, product innovation, and customer service are the other vital factors that play a crucial role in a firm's success.

The competitiveness and ultimate success of a corporation is the result of teamwork that embodies contributions from a large network of resource providers including investors, employees, creditors, consumers and suppliers. The contributions of stakeholders constitute a valuable source of building profitable companies. A stakeholder can be in the form of investor, employee, creditor, supplier or any other person who has interest in a company. Companies in order to be competitive they have to formulate new strategies, which concentrates on stakeholder participation and it may include: employee representation on corporate boards, employee stock ownership plans, profit sharing mechanisms or governance processes that consider stakeholder viewpoints in certain key-decisions e.g. creditors involvement in insolvency proceedings. Consideration of workers as stakeholders in the industrial enterprises is not something new. The idea had its roots in the industrial revolution. As an ideology, employees' participation stems from the belief that a company exercises social and economic function in the structure of a democratic society and evokes in the worker a desire to participate in its affairs.

The concept of worker as a stakeholder contributes to the humanization and democratization of work places. The worker's status as a stakeholder is beneficial to all the parties concerned. The real nature of benefit to each party may however be different. The workers represent human face of an organization. It is they who translate polices laid down by the Board/Management. They are the backbone of an organization, responsible for putting flesh and blood into the organization and are rightfully considered an important stakeholder in the corporation. The table given below provides an overview of various objectives of the stakeholders of a corporate body. (See Table 1)

Table 1: Showing Interests and Objectives of Various Stakeholders

Parties	*Primary Objectives*	*Secondary Objectives*
Employees	Good standard of living	Individual Goals like leading life in a happy and comfortable manner with good job satisfaction
Employer	Improvement in performance	Enhancement of efficiency and maintenance of internal cooperative solutions
State	Social integration	Economic efficiency
General Public	Reduction in living cost and improvement of standard of living	Extended democracy, lower industrial conflicts

An organization needs capital and labour to create wealth. Earlier, the most important need for an organization to be a successful was capital; as long as they had capital, the organization was able to be successful. But today, the need has extended beyond capital and includes labour. The conventional model was the "shareholder capitalism" where the sole emphasis is on strengthening the rights of, and the protection for, financial investors. Today, the growing recognition that human capital is a source of competitive advantage has led to the understanding that labour is, if not, more important at least as important as, capital. Today, corporate leaders in developed countries increasingly understand that people and the knowledge they create are often the most valuable assets in a corporation. This is what they call knowledge capital, which is considered as an invaluable asset of an organization. In fact when a company acquires another company they value human capital more than the plant and machinery. There are a variety of ways by which the interest of employees can be represented in an organization. The growing representation proves that employees' participation does create wealth. There is a need to realize that shareholders' long–run interests are probably well-served by including employees in the formation of building of any company.

5. CORPORATE OBLIGATIONS TO SOCIETY, INVESTORS, AND EMPLOYEES

The corporate bodies' obligations towards various stakeholders are in fact manifold. However, so far no concrete research data is available from this end and accordingly in the present study an effort was made to understand the importance of obligations of corporate bodies and their spirit in fulfilling various obligations towards its customers. In order to understand the scenario a sample of 10 units consisting of 10 manufacturing concerns altogether from Paper, Energy (Thermal – Production) and Pharmaceutical companies in India were considered (The names of the units were not given due to confidentiality). (See Table 2)

The analysis was carried with the help observation, interviews and through a questionnaire to select executives working in CSR, HR and Finance Departments was carried in these units. The observation was carried under three-broad heads:

1. Obligations towards Society
2. Obligations towards Investors
3. Obligations towards Employees.

Table 2: Showing Corporate Obligations to Society, Investors, and Employees

S. No.	Obligations	Fulfilled	Not-Fulfilled	No Data Found
Obligations Towards Society				
1.	Legal Compliances	8	1	1
2.	Corporate Citizenship Behaviour	3	–	7
3.	Ethical Behaviour	2	–	8
4.	CSR	10	–	–
5.	Environment-friendly	1	9	–
6.	Health and Safety Working Conditions	3	6	1
7.	Trusteeship	4	–	6
8.	Accountability	3	2	5
9.	Timely Responsiveness	–	10	1
10.	Political Non-alignment	1	9	–
	Sub Total	*35*	*37*	*29*
Obligations Towards Investors				
11.	Towards Shareholders	8	2	–
12.	Measures to Promote Shareholder Participation	2	6	2
13.	Transparency	–	10	–
14.	Financial Reporting and Records	10	–	–
	Sub-Total	*20*	*18*	*2*
Obligation to Employees				
15.	Fair Employment Practices	2	8	–
16.	Whistle-blowing Encouragement	–	10	–
17.	Human-Treatment	5	5	–
18.	Participation	2	7	1
19.	Empowerment	3	6	1
20.	Equity and Collaborative Environment	1	7	2
	Sub-Total	*13*	*43*	*4*
	Total	**68**	**98**	**35**

Source: Author.

The collected data revealed the following facts:

1. A total of 68 obligations altogether in these 10 units were followed towards society, investors and employees.
2. A total of 98 obligations altogether in these 10 units were not followed towards society, investors and employees.

3. For a total of 35 obligations no data either found or followed in these 10 units.

4. A more or less even numbered obligations were found in terms of fulfilling, not fulfilling and data not found in these select 10 units. Quite interestingly the sub-total obligations towards not-fulfilling and no data found is marginally very high (69) in comparison to obligations fulfilled (35) in these 10 units.

5. Quite interestingly even with respect to obligations towards investors the distribution of score more or less was even in terms of obligations fulfilled and not-fulfilled 20 against to 18.

6. As many as 43 obligations towards employees were not-fulfilled for a total of 10 units against a very low score of 13 obligations, which were fulfilled. It means the concerns towards employees to the maximum extent were not found.

5.1 Outcome of the Analysis

Even though getting competitive advantage is the major focus for majority of the undertakings, but the organizational mission in carrying out of this objective seems may not be possible in the long run due to the following reasons:

1. The scope or importance assigned to societal obligations is inadequate and pawing the same to loose market from the society perspective, if the society perceives the existence of these units is detrimental to the vicinity

2. The major factor for organizational existence is investors, as long as investors obligations were fulfilled as evidenced in earlier researches will lead to better market share and a happy investor further augment and continue the association with the corporate for long-term continuity as a stable stakeholder in the concern

3. Employees are backbone of any concern, they have to be given importance and recognized as vital points for success of any organization, in contrast if the employees are given less recognition and the obligations towards employees if ignored can lead in the long run losing of talent and it can have a quite bearing on its long-term plan of gaining competitive advantage. In fact, it is the employees, who will march towards the objective of becoming a competitive organization, if they are satisfied.

6. SUSTAINABILITY

Organizations major thrust in any time zone is to meet their own needs without jeopardizing the society and natural environment. The magnitude of concentration towards society and environment is increasing day by day. Organizations, which intend to become good corporate citizen, are incorporating in their long term plans, strategies for dealing with societal needs, natural environment and corresponding business imperatives. Organizations are finding it difficult to concentrate on these three aspects. In order to balance these aspects in an optimal way, a framework they thought of to bring. One such framework is Triple Bottom Line. TBL is in fact a step towards increasing organization vis-a-vis environmental sustainability.

The concept of sustainability urged nations to find alternative means of economic expansion without destroying the environmental resources or sacrificing the wellbeing of future generations. Sustainability, since then become a buzzword in corporate mainstream agenda.

With respect to TBL, which is a part of sustainability reporting was mooted by **John Elkington**, who first coined the term and propagated it across the world in order to educate the people about the three way reporting framework for People, Planet and Profit, that goes beyond the boundaries of traditional reporting practices. According to him, TBL focus on:

1. Economic Sustainability
2. Environmental Sustainability
3. Social Sustainability.

In fact, sustainability requires looking holistically at the various elements of business operation that determines business success. These elements include business profitability, and protecting the environment for future generations through environmental performance. Triple bottom line is basically an accounting framework, which incorporates three dimensions of performance. They are:

1. Social
2. Environmental, and
3. Financial.

These three dimensions are commonly called as People, Planet and Profit as portrayed in the Figure 2.

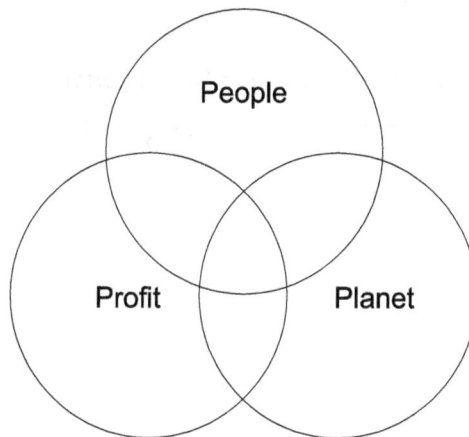

Fig. 2: Framework of TBL

6.1 Modus Operandi for Environmental Sustainability in India

Environmental Sustainability is one of the key requirements for business continuance. Degradation of environment dilutes the very existence of human. Therefore, virtual Governments should plan for environmental protection. One such exercise is implementation of strict Corporate Social Responsibility rules. The concept of CSR is widely acknowledged in most of the developed economies. However, in the case of developing and under developed economies Governmental intervention is highly required. For example in the case of India, the Government of India adopted in The Companies Act, 2013 the concept of "Corporate Social Responsibility" (CSR). As per Sec. 135 of the Companies Act, 2013 companies with a net worth of ₹ 500 crore or more; a turnover of ₹ 1,000 crore or more; and a net profit of ₹ 5 crore or more during any financial year are compulsorily required to follow CSR provisions in India.

6.2 Advantages of Sustainability to Business

1. Reduce operating costs by undertaking initiatives that reduce waste, water and energy Consumption.
2. Develop a competitive advantage by establishing and promoting sustainable practices as a Point of difference.
3. Attract and retain valuable staff by adopting policies that meet with employee values and concerns. Also, by creating an environment of team knowledge sharing, best practice, and innovative ideas, employees will feel better about their work environment, and new employees may be attracted to the business.
4. Encourage investors interested in companies with long-term sustainability plans that minimize operating risks in the future.
5. Increase long-term profitability by maximizing your business potential and putting plans in place now that will create savings in the future.
6. Know that your actions are actually making a difference to protect and enhance natural resources for future generation.

6.3 Is Sustainability a Preach or Practice in Organizations?

Sustainability in order to be seen in organization, the organization concerned should exhibit certain modalities about its concern over sustainability. There some yardsticks, which can measure organizations real spirit on sustainability, focus. Some of the yardsticks are tabulated (See Table 3) as follows:

Table 3: Sustainability Practices in Organizations

S. No.	Sustainability Practices/Yardsticks	% of Share Out of 100
1.	Recycling	
2.	Community Investment	
3.	Energy Conservation	
4.	Community Educational Programmes	
5.	Waste Reduction	
6.	Fair Trade Practices	
7.	Pollution Reduction	
8.	Alternative Energy Use	
9.	Any Other Specific Measure	

Source: Author.

Virtual organizations should be able to demarcate the % of contribution towards sustainability. Organizations have to answer, whether these practices at all are adopted or not? It not adopted, they can certainly come under Preach type. If they practice, they have to express these parameters in quantitative dimension.

The above quantification is essential for the very survival of that industry only. If it ignores these measures sooner or later it will lose its competitive advantage.

7. LIMITATIONS OF THE STUDY AND SCOPE FOR FUTURE RESEARCH

The research has certain limitations like by not taking into account size of the units and annul turnover, the current research is more or less is confidential in nature and the respondents are not interested to share each and every aspect in spite of assurance of non-revealing of results and another limitation is non-adoption of standard metric to determine the validity of results as it is purely a conceptual research.

8. SUGGESTIONS FOR GOOD CORPORATE GOVERNANCE

The current research felt a need to streamline corporate governance and accordingly it came out with the following recommendations:

1. The investors' confidence is the most important criterion as long as they have confidence in financial and capital market, the economy of India progress and this call for a great amount of transparency, which is possible only with the help of a good corporate governance system.
2. Transparency can be promoted through good corporate procedures, disclosure and accountability.
3. Recognize employees, manager and board members who are ethical and percolate good governance in the organizations. It is through people good governance can be achieved and not through regulations. People have to be appraised on value systems in all decision making activities.
4. Employee should know much early in advance the consequences of non-compliance, so that the principle enunciated by Henry Fayol (esprit de corps) can be made it possible really.
5. With respect to whistle blower policy, the organizations have to make it compulsory to follow on their own initiative, though the apex bodies are silent on this policy being it is non-mandatory in nature. Organizations have to make attempts to promote ethical behaviours in their culture. In order to avoid a situation like this nature, they can choose value systems that govern the values of both employees and employer. Organizations should check before making any claims about its possibility for implementation.
6. The practice of ethical testing must be employed without any bias to all employees and the results must be properly analysed to know the drawbacks. This sort of system acts as a check and identifies its importance.
7. There is a need for Indian companies to adopt compulsory "Ethics Training". There is a need that "ethics not only to be taught but also it has to implemented" in its true spirit.

8. Before adjudicating any award to corporates, the concerned bodies declaring award must take thorough scrutiny in all aspects. Otherwise, not only the sanctity of the award but also the concerned organization's reputation will be seriously defamed.

9. SCOPE FOR FUTURE RESEARCH

Corporate governance has received much attention in recent years, partly due to the financial crisis throughout the world. A review of the literature on corporate governance issues confirms that, corporate social responsibility is one of the major determinants of organizational existence in future. Therefore, concrete research in providing exact amount or figure to be contributed by each and every corporate body has to be earmarked through concrete research.

TBL reporting has given lot of scope of for future research, a concrete research to be carried on with the help of general financial reporting.

Corporate Governance could not exist without human intervention as long as there is ethical and prudent behavior on the part of employees it cannot be a hard nut to crack, therefore concrete research on employee governance is more required.

The future research should be directed towards determining the ideal share between people, planet and profit for any organization.

10. CONCLUSIONS

Corporate Governance framework though gained its importance but in terms of fulfilling obligations, especially societal and employee point of view it has to march towards new horizons because organizations are still more or less investor oriented and investors to the maximum extent may oppose practices and policies towards the well-being of the society as well as towards employees, if the trend is changed, there could be definitely a good corporate governance the corporate world can see. Therefore, securing good corporate governance is quite challenging.

Further, the adoption of proper sustainability requires increased participation of employees and top level management in the organization. This can be well achieved through a good TBL (Triple Bottom Line) reporting. In one way though the organizations should have the perspective of wealth creation, it should not be at the expense of the benefits and wages due to the employees.

To sum up even though the adoption and implementation of good corporate governance initiatives and Triple Bottom Line reporting is complex, it is the organizations and their will that allows the organizations to apply these concepts and frameworks in a manner suiting their requirements of all the stakeholders. In fact, good corporate governance cannot be brought through papers it is the will of all stakeholders, who can only bring good corporate governance.

REFERENCES

Andrew, Savitz (2006). *The Triple Bottom Line* (San Francisco: Jossey-Bass, 2006).

Asha Bhandarkar, Journal Article: Making People Work Millennials and the Workplace: Challenges for building the Organization of the future, Indian Management, October, 2012, Vol. 51, Issue 10, p. 58.

Charles, R. Greer (2012). Strategic Human Resource Management, Pearson, pp. 141–170.

Chatterji, Madhumita (2011). "Corporate Social Responsibility", Oxford Higher Education, Oxford University Press, pp. 1–57.

Chiplin, B. and Coyne, J. (1977). Can Workers Manage? Part I: Property Rights, Industrial Democracy and the Bullock Report, The Institute of Economic Affairs, pp. 22–32.

David Lepak and Mary Gowan (2009). Human Resource Management Managing Employees for the Competitive Advantage, Pearson, pp. 25–50.

Dev, Chatterjee (2013). "What you need to know about the new Company Law and CSR" Business Standard August 09, 2013.

Donaldson, L. and Davis, J.H. (1919). 'Stewardship Theory or Agency Theory: CEO Governance and Shareholder Returns', Australian Journal of Management, Vol. 16(1).

Dutta Sumanta (2011). Triple Bottom Line Reporting: An Innovative Accounting Initiative *International Journal on Business, Strategy and Management*, Vol. 1, No. 1, June, 2011, pp. 1–13.

Fernando, A.C. (2009). "Corporate Governance: Principles, Policies and Practices", Pearson, pp. 3–68.

Frank, Vancley (2004). "The TBL Impact Assessment: How do EIA, SIA, SEA and EMS Relate to Each Other? *Journal of Environmental Assessment and Policy*, Vol. 6, Issue 3. http://www.worldscientific.com/doi/abs/10.1142/S1464333204001729.

Gary Dessler and Biju Varkkey (2011) "Human Resource Management", Pearson, pp. 84–113.

Hubbard, Graham (2009). "Measuring Organizational Performance. Beyond the Triple Bottom Line" Bus, strategy and the Environment, 19, 177–191.

Irani, J.J., Raha Subir and Prabhu Suresh (2005): "Corporate Governance: Three Views" Viklpa, Vol. 30, No. 4, October–December, 2005.

Jackson, I.A. and Nelson, J. (2004). Profits with principles: Seven strategies for delivering value with values. New York, NY: Doubleday.

John Elkington, "Towards the Sustainable Corporation: Win-Win-Win Business Strategies for Sustainable Development," *California Management Review* 36, No. 2 (1994), pp. 90–100.

Jyothi, P. and Venkatesh, D.N. (2011). "Human Resource Management" Oxford University Press, pp. 19–30.

Li-Chin, Jennifer Ho and Martin, E. Taylor (2007). "An Empirical Analysis of TBL Reporting and its Determinants: Evidence from US and Japan", *Journal International Financial Management and Accounting*, Vol. 18, No. 2, pp. 123–150.

Miles, Raymond E. and Charles, C. Snow, "Designating Strategic Human Resources Systems," Organizational Dynamics 13, No. 1 (1984), pp. 36–52.

Murthy, V. and Abeysekera, I., Corporate Social Reporting Practices of Top Indian Software Firms, Australasia Accounting Business and Finance Journal, 2(1), 2008.

Nancy Fell, "Triple Bottom Line Approach Growing in Nonprofit Sector, "*Causeplanet*, January 21, 2007, and Peter Senge, *et al., The Necessary Revolution* (New York: Doubleday, 2008).

Porter, Michael E., Competitive Advantage Creating and Sustaining Superior Performance, New York: The Free Press, 1985.

Rao, V.S.P. (2009). Human Resource Management, Excell Books.

Ruth, Aguilera and Geroge, Y.I.P., "Global strategy faces local constraints" Article published in Financial Times, 27 May, 2005.

Stephen, R.J., Sheppard and Michael, Meitner, "Using Multi-Criteria Analysis and Visualization for Sustainable Forest Management Planning with Stakeholder Groups," *Forest Ecology and Management* 207 (2005): 171–187. Another example can be found in Katrina Brown *et al.*, "Trade-Off Analysis for Marine Protected Area Management," *Ecological Economics* 37, No. 3 (June 2001), pp. 417–434.

Sustainable Cleveland, 2019, "Action and Resources Guide: Building an Economic Engine to Empower a Green City on a Blue Lake, "October, 2010 www.gcbl.org/system/files/SC2019+ Executive+Summary+%289SEP10%29.pdf.

Tyleca, D., CarLens, J., Bechhout, F., Hertin, J., Wehrmeyer, W. and Wagner, M. (2002). "Corporate Environmental Performance Evaluation: evidence from the MEPI project, Business Strategy and the Environment II, pp. 1–13.

Vanek, J. (ed.) 1975, Self-Management: Economic Liberalization of Man, Penguin, pp. 16–17.

A Study on Eco-friendly Paper Products Based on Sustainable Design Principles

P. Deepasri

Department of Fashion Design, Vogue Institute of Fashion Technology, Bengaluru, India

ABSTRACT: *The technological advancement and sophistication of lifestyle has lead to a Plastic world which are non-biodegradable, toxic and hazardous to the environment. These products have failed to establish emotional relationship with the user, which is one of the core concept of Sustainable product development. Therefore, the study aimed at designing and development of sustainable eco-friendly paper products which are manufactured using textile fibers like cotton, silk and other cellulosic fibers, with specific utility feature. It was based on "Sustainable Design Principles", which made use of locally available low-impact materials like handmade papers, recycled boards, natural colors and pigments and other non-toxic materials. The energy consumed was low and required less stages of processing. Emotionally durable designs were achieved by implementing the elements of Design in an innovative way. The concept of reuse and recycling made the products sustainable and eco-friendly. The products were designed with a specific end-use and attractive color paper range was used to bring out a collection. Traditional motifs of Madhubani, Warli, Pat Chithra, Chukki-Chithra, Kantha, Banjara and other tribal inspired motifs were used to decorate the surface of the paper products. A natural and textile feel was simulated to enhance the aesthetic appearance of the product which is also its USP. The production process was carried out using basic tools and manual methods of marking, folding, creasing, cutting and assembling was used. The motifs were carefully chosen to suit the contour and the size of the product. Varying color media were used supporting different textures of paper. Utility products like office stationary, jewelry boxes, lamp shades, tea coasters, paper bags, paper pens, vases, dust bins, etc., were made with an edge over the quality. The products were attractively packed and 100 urban customers, from various professions within the age group of 25–30 were selected. The respondents were given the products and were asked to use the products for a week's time, subjecting to "real environment" testing and feedback was taken. The ergonomics of design was assessed based on the test. At a reasonable cost aesthetically rich and personalized products were made available to the customers, which replaced the mediocre plastic made products. The usage of handmade papers made of textiles fibers added rich texture to the products. The paper products were rated versus plastic products based on the core features of the products like eco-friendliness, simplicity of manufacture, usage of locally available raw materials, innovation in design feature contributing to high aesthetic value, when compared to plastic utility products, the ability of reuse and recycle, ease in maintenance, ergonomic factors while deciding the geometry and space in construction of the products, thematic selection of motif, unique design composition and repeats and room for customizing product features with little variation in process of making. Though the products were targeted to a niche market, it has the potential of becoming a specific product sector in the future for corporate gifts, office utilities, home-décor and custom* *made signature product line. This can also be a skill enhancement cluster program which will help to generate skilled and small entrepreneurs locally, thereby contributing to the economy of their community. Further, this area can be researched for innovation.*

1. INTRODUCTION

The technological advancement and sophistication of lifestyle has lead to a Plastic world which are non-biodegradable, toxic and hazardous to the environment. These products have failed to establish emotional relationship with the user, which is one of the core concept of Sustainable product development. Therefore, the study aimed at designing and development of sustainable eco-friendly paper products which are manufactured using textile fibers like cotton, silk and other cellulosic fibers, with specific utility feature. It was based on "Sustainable Design Principles", which made use of locally available low-impact materials like handmade papers, recycled boards, natural colors and pigments and other non-toxic materials. The energy consumed was low and required less stages of processing. Emotionally durable designs were achieved by implementing the elements of Design in an innovative way. The concept of reuse and recycling made the products sustainable and eco-friendly. The products were designed with a specific end-use and attractive color paper range was used to bring out a collection. Traditional motifs of Madhubani, Warli, Pat Chithra, Chukki-Chithra, Kantha, Banjara and other tribal inspired motifs were used to decorate the surface of the paper products. A natural and textile feel was simulated to enhance the aesthetic appearance of the product which is also its USP. This paper specifically aims at identifying the suitable design practices and product concept which can accommodate traditional motifs thereby enriching the product value distinctly.

2. METHODOLOGY

2.1 Raw Materials and Tools Used

The key raw materials that were used for the product development were:

1. *Papers*—Recycled papers, recycled boards, liners, handmade papers, linen boards, oil paper, tracing sheets, transparent sheets, hybrid papers (manufactured partly by hand and by automation using textile fibers like silk, cotton fibers, etc.).
2. *Basic tools*—Scales, scissors, cutter (of varying widths), bodkins, eyelets, racing wheel, creasing tongue, metal clippers, grippers, metal clips, strings, weights, etc.
3. *Auxiliaries*—Dye (natural and direct dyes), salt, coating agent (natural), gum paste, bonding agents and adhesives.
4. *Paints and colors*—Acrylic colors, binders, water based colors, pigments, markers, colored pens, crayons, etc.
5. *Ornamentation materials*—Kundan stones, beads, ribbons, fabric appliqués, laces.

The broad areas of methodology include.

Step 1: Market Exploration

To achieve the set objectives of the study, product market was explored to collect information about the materials used, the design and to understand the prevailing manufacture and design practices being followed. Also, an extensive exploration was made in the areas of eco-friendly and sustainable product development to understand the intricacies involved. Market exploration helped to find out the trending products and their features in terms of utility and aesthetics. With these findings, a concept for eco-friendly paper products with specific utility features was designed. In accordance with the product development planning, necessary data were collected in the area of traditional motifs, types of papers available in the market, methods that can be used to apply the motifs on the surface of the products. Local sourcing for raw materials and supplies enabled a quick response for the material needed during the stages of product development.

Step 2: Concept Development

The concept of developing eco-friendly products was based on the components of sustainable design development involving the usage of low-impact materials, which are non-toxic, sustainably produced, recycled which require low energy to process. Emotionally durable design—by increased relation between end-users and products, with effective designing. The reuse and recyclability of products was also one of the main concerns. The concept of 'Paper Utility products' was combined with an idea of drawing and painting the traditional Indian motifs on the surface of the products. Therefore, to support the concept, the texture of the paper and the color schemes were carefully selected. The selection of the motifs was done based on the nature of the product, contour and the surface. Suitable color schemes and color media were chosen. A theme was chosen to coordinate the office and home utilities specially designed for men and women distinctly. The shape and size of the products were designed uniquely and the motifs were selected to suit the overall product design. The products were designed based on ergonomics of usage to facilitate smooth handling and use. A theme board was created to compile and present the factors influencing the collection. It included the selection of papers, color schemes, motifs and expressions, all drawn from the theme. A practical approach to product design and development was done by illustrating the product designs in the sketch book to arrive at a range. Care was taken to make sure that the concept of space was the focal point. All the usage aspects within a product was paid attention and the parts were designed in such a way that it supports the product to be complete and usable without wasting any space or material. Final selection of designs was done to bring out a range out of it.

Step 3: Range Development

The range included office table-top utilities like slip pads, writing pads, pen stand, trays, paper weights and home decor items like frames, vases, lamp shades, serving trays, tea coaster sets, etc. The approach for prototyping involved interaction with the customers to understand and to arrive at the usability factor of the products. This facilitated the discussions with the users. The usefulness of the products designed and new requirements for functionality were uncovered during the discussion, which helped to gradually improve the usability of the products. Also, the application of traditional motifs on the product surface was appreciated. These were illustrated on a sketchbook to start off with production.

Step 4: Prototyping

The Design prototypes were done in order to assess the form, fit and usability of the product range designed. It included both the functional and aesthetic properties of the product. The products were manufactured using manual methods with simple hand skills rather than automated systems. A set of basic hand tools were used to make the products, which involved the stages of measuring, cutting, shaping, creasing, gluing, flattening, followed by drying. After drying, where ever needed surface coating was done using an eco-friendly corn based protein coating agent which was locally available. The surface of the products was decorated by applying traditional designs and motifs. It was accomplished by hand drawing and painting using varying techniques and color media. The type of motif, its size, repeat and composition was paid at most importance. The painted motifs were enriched by using kundans, metallic paints and novelty beads and metallic sequins.

The stages involved in the process of product making:

1. *Selection of product design for production*—Around 10 designs were selected to be converted into products. The selection was based on the utility features of the design.
2. *Selection of paper*—Paper being the prime raw material, care was taken while selecting the papers. Paper with compatible surface for surface painting and value addition were selected.
3. *Selection of the motif*—The motifs to be applied on the product surface were selected based on their shape, purpose, the area to be applied and the overall proportion with the product.
4. *Making the paper product*—The required measurements for a product were taken and templates for the same were created using recycled papers. The shape and space needed was considered while including the allowance for gluing and bonding. the components of the product were cut using scissors and cutters and the assembling was done in the following stages:
   - *Marking*—It involved the marking of margin, for folding and the lines for creasing.
   - *Creasing*—Creasing was done manually. The paper was folded along the line meant for creasing.
   - *Gluing and bonding*—The components which were required to be bonded together were glued or bonded using gum or synthetic bonding agents.
   - *Flattening*—The process of flattening was carried out when the glue is still wet to avoid occurring of air bubbles.
   - *Drying*—The drying is done naturally, direct sunlight is avoided which can cause fading of the color.
5. *Application of motif*—The selected motifs were applied on the product surface in the following stages:
   - *Preparing the surface*—The surface of the product was cleaned by wiping it with a moist cloth to remove dust and dirt particles.
   - *Drawing/tracing*—The motif was traced or drawn on the surface of the product.
   - *Coloring*—The technique of coloring was decided based on the nature of the motif. It included both brush and non-brush techniques.

6. *Finishing*—The painted surface of the product was finished by coating it with a natural coating agent, extracted by corn. The corn protein gave a remarkable luster to the painted surface.

7. *Drying*—After coating, the products were allowed to dry in natural condition for 24 hours.

8. *Attachments*—Depending on the utility feature, the products were attached with suitable accessories like clips, eyelets, velcro, buttons, strings, etc.

3. RESULTS AND DISCUSSION

The core features of the products were eco-friendliness, simplicity of manufacture, usage of locally available raw materials, innovation in design feature contributing to high aesthetic value, when compared to plastic utility products, the ability of reuse and recycle, ease in maintenance, ergonomic factors while deciding the geometry and space in construction of the products, thematic selection of motif, unique design composition and repeats and room for customizing product features with little variation in process of making.

The features of the product range are as follows:

1. *Paper Tray and Mat*—The tray was made using fused satin paper, by interlacing techniques. The strips of paper were cut and interlaced. The mat was also constructed using the same technique. It can be easily cleaned and maintained. It can also be used as a serving tray.

Fig. 1: Paper Tray—Interlacing Technique

2. *Serving Tray*—A fabric design simulation with traditional paisleys has been used by covering the basic tray structure with the shell paper by bonding technique. It has got a rich look because of the paisleys and their composition.

Fig. 2: Serving Tray in Traditional Paisley (Bonding Technique)

3. *Mirror Frame*—A artistic mirror frame, highlighted with simple strokes techniques to create a traditional border. Aesthetically rich and ease in maintenance.

Fig. 3: Mirror Frame—Stroke Technique of Painting

4. *Gift Envelopes*—Created using recycled and handmade paper. Simple line sketches and Warli motifs were used for the surface decoration.

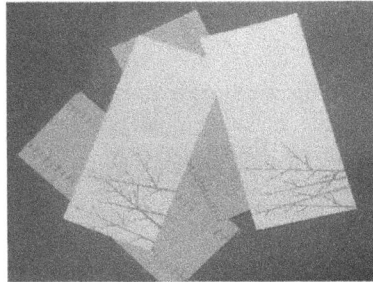

Fig. 4: Gift Envelopes—Line Sketches and Warli

5. *Paper Bags*—Were made in papers of higher GSM, without grain line and a high degree of flexibility. These were the gift bags, simple yet sturdy, surface decoration done as per the concept needed.

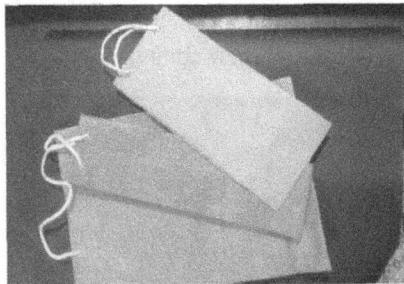

Fig. 5: Paper Bags with Customized Surface Work

6. *Paper Pens*—Made of handmade papers, news papers, decorated with quilling beads and surface designed with traditional motifs.

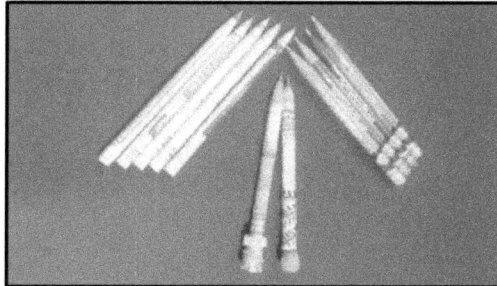

Fig. 6: Paper Pens-Creative Designs

7. *Vase*—A flower vase as a home decor product designed with handmade paper and paisley on surface.

Fig. 7: Vase-Home Decor

8. *Photo Frame*—Designed in rich satin brocade paper, simulation of textile design.

Fig. 8: Photo Frame in Brocade Paper

9. *Dustbin*—Made out of recycled board with a handmade paper covering. It is decorated with a temple border and done in traditional textile color combination of maroon and mustard yellow. Meant for dry paper wastes generated in offices.

Fig. 9: Dustbin with Temple Border

10. *Table Top Utility Set*—Designed with a theme for men, a set of stationery including pen stands, slip box, visiting card holder. It was based on traditional art of Chukki-Chithra.

Fig. 10: Table Top Utility with Chukki-Chithra

11. *Office Set*—Designed for women with the simulation of Varanasi silk brocade.

Fig . 11: Office Set with Varanasi Brocade Simulation

12. *Paper Weight*—Made with cardboard pieces wrapped with a textures covering paper. It is decorated with Rajasthali-Morni motif.

Fig. 12: Paper Weight Designed with Rajasthali Morni Motif

13. *Writing Pad*—Designed with madhubani peacock motif and tribal floral motifs.

Fig. 13: Writing Pad Decorated with Madhubani Peacock Motif and Floral Motif

The prototypes created were tested for usability. A 'real—environment' was chosen for the testing of products, where the products were subjected to usage in the same way it was perceived to be used. A responsive approach was designed to receive responses from the end users to assess the quality and functionality of the products. The end users were approached with the products and direct response was documented. The responses collected as data from the end-users were subjected to analysis of eco-friendly products versus plastic products.

The prototypes created were tested for usability. A 'real–environment' was chosen for the testing of products, where the products were subjected to usage in the same way it was perceived to be used. A responsive approach was designed to receive responses from the end users to assess the quality and functionality of the products. Hundred end users were approached with the products and direct response was documented. The responses collected as data from the end-users were subjected to analysis of eco-friendly products versus plastic products. The paper products were rated versus plastic products on the parameters of Functionality, Aesthetics, Innovation, Variety, Surface Design Appeal, Geometry of Construction, Space, Eco-friendliness, Maintenance, Accessibility, Quality and Cost-effectiveness.

Paper Products Versus Plastic Products

Number of Respondents–100

Parameters	Paper Products			Plastic Products		
	Average	Good	Excellent	Average	Good	Excellent
Functionality		30%	70%	60%	20%	20%
Aesthetics			20%	70%	30%	
Innovation		40%	60%	90%	10%	
Variety		20%	80%	18%	90%	
Surface Design Appeal			100%	90%	10%	
Geometry of Construction		10%	90%	60%	40%	
Space		10%	90%	10%	75%	15%
Eco-friendliness			100%	0	0	0
Maintenance	10%	70%	20%	10%	90%	
Accessibility	60%	25%	15%		10%	90%
Quality		10%	90%	40%	50%	10%
Cost-effectiveness		20%	80%	40%	10%	50%

The paper products were found to be more competitive in comparison with plastic products. Usage of paper as the prime raw material enhanced the aesthetic value of the products by adding high textural value to it. The color schemes used made the products more eye catchy and impressive. The strong structure and geometry used in the products made them unique and the application of traditional motifs made the products distinctive, classic and valued for money. The reuse–recyclability of products made it much preferable than the plastic products. All these factors contributed to the higher acceptance of paper products than plastic products.

4. CONCLUSION

The 'sustainable design principle' was proved to be an effective method to develop eco-friendly paper product line, for a concept of utility product design. The usage of locally available materials, manual methods of production contributed to support high product variation and economy of manufacture. The design testing gave a positive outcome and the acceptability test was conducted in a real environment and the target end-users gave a remarkable feedback indicating the demand and potentiality for such products in the urban market. The experimentation of applying traditional Indian motifs as surface designs was well accepted and an attempt to popularize the traditional Indian motifs by bridging the gap between 'art and man' in today's sophisticated lifestyle was accomplished to some extent through the product development. The high aesthetic value of the products became one of the major USP. From the study, it was convincing that urban customers accept and appreciate traditional motifs with an edge of utility. It was also evident that Paper utilities can be a good substitute for plastic utilities. This pilot research can be further explored and experimented on the similar lines of design and product development.

REFERENCES

Berkowitz, Marvin (1987). "The Influence of Shape on Product Preferences," Advances in Consumer Research, eds. Melanie Wallendorf and Paul Anderson, *Provo UT: Association for Consumer Research,* Vol. 14, p. 559.

Caniato, Federico; Caridi, Maria; Crippa, Luca and Moretto, Antonella, Environmental sustainability in fashion supply chains: An exploratory case based research, *International Journal of Production Economics,* Vol. 135, Issue 2, February 2012, pp. 659–670.

Geetha, D., Jenifer and Annie, D., A study on Consumer Behaviour towards Purchase of Eco-friendly products in Coimbatore, *Abhinav International Monthly Refereed Journal of Research in Management and Technology,* Vol. 3, Issue 3 (March, 2014) Online ISSN-2320-0073.

Holbrook, Morris B. and Zirlin, Robert B. (1985), "Artistic Creation, Artworks, and Aesthetic Appreciation: Some Philosophical Contributions to Nonprofit Marketing," *Advances in Nonprofit Marketing,* Vol. 1, pp. 1–54.

Khan, M. Adil, Sustainable development: The key concepts, issues and implications, published by John Wiley and Sons, Vol. 3, Issue 2, pp. 63–69, 1995.

Lee, Kaman (2008). "Opportunities for green marketing: Young consumers", Marketing Intelligence and Planning, Vol. 26, Iss. 6, pp. 573–586.

Olson, Jerry C. (1981). "What is an Esthetic Response?" Symbolic Consumer Behavior, eds. Elizabeth C. Hirschman and Morris B. Holbrook, Ann Arbor, MI: Association for Consumer Research, pp. 71–74.

Paul. M. Bator, An essay on the International Trade in Art, published by Stanford University, Vol. 34, No. 2, Jan 1982, pp. 275–384.

Punam, Rani and Vivek, Singh, Traditional to Contemporary Indian Jewellery: A Review, Global research analysis, Vol. 2, Issue 1, Jan. 2013, ISSN No. 2277-8160.

Robert, W. and Veryzer, Jr. (1993). "Aesthetic Response and the Influence of Design Principles on Product Preferences", in Advances in Consumer Research, Vol. 20, eds. Leigh McAlister and Michael L. Rothschild, Provo, UT: Association for Consumer Research, pp. 224–228.

Schot, Johan and Geels, Frank W., *The Dynamics of Sustainable Innovation Journeys,* Vol. 20, Issue 5, 2008.

Shaun, McNiff, Art based research, handbook, Knowles Publications 2007.

Treadaway, Cathy, The Impact of Digital Imaging Technology on the Creative Practice of Printed Textile and Surface Pattern Design, North Carolina State University, 2004.

Conditional Value-at-Risk Based Portfolio Optimization Using Particle Swarm Optimization

Jhuma Ray[2] and Siddhartha Bhattacharyya[2]

[1]Department of Science and Humanities, RCC Institute of Information Technology,
Canal South Road, Beliaghata, Kolkata–700 015, India
[2]Department of Information Technology, RCC Institute of Information Technology,
Canal South Road, Beliaghata, Kolkata–700 015, India
E-mail: [1]jhuma706@gmail.com; [2]dr.siddhartha.bhattacharyya@gmail.com

ABSTRACT: *In presence of volatility in today's real work-a-day world's financial transactions, an equitable balance between risks and returns has to be maintained by any investor to derive at an optimum standpoint. In spite of the prevailing volatility, the boon in disguise lies in the correlation of the combination of financial instruments/assets in a financial portfolio in a particular market condition. Of late, portfolio management has been necessitated due to the need for decision making in investment opportunities in a high-risk scenario. It addresses the risk-reward tradeoff in allocation of investments to a number of different assets so as to maximize returns or minimize risks in a given investment period. This article delineates an algorithm of particle swarm optimization accompanied by optimized portfolio asset allocations in a volatile market condition. The proposed approach is centered around optimizing the Conditional Value-at-Risk (CVaR) measure in different market conditions based on several objectives and constraints. The results are compared with those obtained with the optimization of Value-at-Risk (VaR) measure of the portfolios under consideration. A comparative application of the proposed approach along with the VaR approach is demonstrated on a collection of several financial instruments.*

Keywords: Portfolio Management, Risk-return Paradigm, Value-at-Risk, Conditional Value-at-risk, Particle Swarm Optimization.

1. INTRODUCTION

In presence of volatility in today's real work-a-day world's financial transactions, an equitable balance between risks and returns has to be maintained by any investor to derive at an optimum standpoint (Brown, 2004; McNeil *et al.*, 2005). In spite of the prevailing volatility, the boon in disguise lies in the correlation of the combination of financial instruments/assets in a financial portfolio in a particular market condition. Of late, portfolio management has been necessitated due to the need for decision making in investment opportunities in a high-risk scenario hence proving the present day's scenario risks and returns to be inevitably interlinked resulting in the importance in the decision making procedure in the investment opportunities. It addresses the risk-reward tradeoff in allocation of investments to a number of different assets so as to maximize returns or minimize risks in a given investment period. According to

Markowitz (1952) (Markowitz, 1952) selection of an asset should not be done depending only on its characteristic features but also taking into account its co-movement with other assets. Computation of risk as standard deviation of returns was done by Markowitz also showing diversification into different investment factors which in turn have limited or negative correlations in terms of their movements reducing overall risk. This movement is measurable by a correlation coefficient varying between +1 and −1 according to Markowitz.

There have been several models for portfolio selection which have come up throughout the years. These include the early mean variance models subject to Markowitz's work in 1952 (Markowitz, 1952).Of late, a host of stochastic optimization methods based on the market scenario have assumed importance (Hiller and Eckstein, 1993; Birgeand Rosa, 1995; Mulvey *et al.*, 1995; Bai *et al.*, 1997; Vladimirou and Zenios, 1997; Cariño and Ziemba, 1998). No matter which model one resorts to, the underlying principle/notion lies in the denigration of some measures of market risks coupled with focus on increasing the portfolio return. It has been noted that the risk metric is simulated to be a function of the possible portfolio returns in almost every models.

The most widely used approach for gauging the negative aspect of risk in a portfolio is the Value-at-Risk (*VaR*) which is designated as the pth percentile of portfolio return at the edge of the planning perspective. Incidentally, for low values of p (as low as 1, 5 or 10) it identifies the "unconditional" fallout of portfolio returns. Stambaugh (1996) envisages *VaR* to be 1) a terminology for risk 2) giving space to efficient and coherent risk management 3) administering an enterprise-wide technique for market regulation and 4) acting as a devise for risk judgement.

A score of literature exists regarding the varied techniques for the computation/calculation of a portfolio's *VaR*. One of the appealing perspectives of *VaR* derivation is to find out as to how it can be applied for allocation of portfolios in multi-financial instruments situation. If business organizations are to perform based on *VaR*, then it is a vital issue in designing the strategy for investment selection. Furthermore, if organizations take decisions in a *VaR* context, then the implications of the organization's risks are to be taken into cognizance. Since *VaR*is rather discrete in nature and is strenous to assimilate in conventional stochastic models, not much works have been reported in the literature as regards to its optimization attempts. Rockafellar and Uryasev (2002) proposed a scenario-based model for portfolio optimization. He adopted the Conditional Value at Risk (*CVaR*) for this purpose. *CVaR* is exemplified as the anticipated value of losses outstripping VaR. Their model lessens *CVaR* in course of scheming *VaR*. It was observed that the minimum-*CVaR* is tantamount to the minimum-*VaR* in the case of normally distributed portfolio returns.

Thus for measuring the value of an asset or of a portfolio of assets in the market which get decreased over a certain time period (usually considered over 1 day or 10 days) subservient to typical market conditions, Value-at-Risk (*VaR*) (Dowd, 2005; Holton, 2003; Jorionand Philippe, 2001) stands to be an effective tool. It is also highly valued for being incorporated within industry regulations (Jorion, 2001), regardless suffering from the unstableness as well as difficulty to work using numerical values in case of normal distribution of losses because loss distribution often tends to display "fat tails" or factual discreteness.

Value-at-Risk (*VaR*) (Jorion, 2001) at a confidence level is thus considered to be the maximum loss not exceeding with a given probability, over a stipulated time period. *VaR* is always been determined by three parameters, *viz.* (i) the time horizon (typically 1 day, 10 days, or 1 year) which is to be analyzed as the time over which any organization should hold its portfolio, or to the time required for liquidating its assets, (ii) the confidence level (common values are 99% and 95%), which is the estimate of the interval where the *VaR* would not likely exceed the unit of *VaR* in currency and (iii) the maximum probable loss structure.

Unlikely Value–at-risk, Conditional value-at-risk (*CVaR*) stands to be a risk measuring technique in case of risk having significant advantages, for deriving distribution of losses in finance involving discreetness (Rockafellar and Uryasev, 2002). Because of the customariness of different proposed structures found onvaried scenarios and finite sampling, the application of such distributions have become an important property in the financial markets in turn.

CVaR can be identified to be the weighted average of *VaR* along with $CVaR^+$ (the values themselves be contingent on the decision x along with the weights), where none of the values of *VaR* and *CVaR+* stands to be coherent. The particular way of computing of *CVaR* in terms of probability of *VaR* value, gives birth to the value of weights, when one exists.

Computational advantages of *CVaR* over *VaR* have become the major impetus in the *CVaR* methodology development procedure, in spite of substantial efforts for finding out the efficient algorithms for the process of optimization of *VaR* in high-dimensional environments which are still unavailable. *CVaR* stands to be a new coherent risk measuring structure having distinct advantages when been compared to *VaR* (Rockafellar and Uryasev, 2002), quantifying risks beyond *VaR*, consistent at different levels of confidence α (smooth with respect to α) and also being a static statistical estimate with integral characteristics. *CVaR* is thus been entrenched to be an excellent tool in the risk management procedure and optimization of portfolio accompanied by linear programming having huge dimensions in the company of substantial numerical implementations. At various time periods with different levels of confidence, distributions are also shaped for multiple risk constraints along with the previously mentioned tasks, which in turn stand as fast algorithms for online usage. Rockafellar and Uryasev (2002) (Rockafellar and Uryasev, 2002) have considered *CVaR* methodology to be a consistent one having mean-variance method taken to be under minimal portfolio (with return constraint), which can also be considered to be a variance minimal in case of normal loss distribution.

In this article, an algorithm exercising particle swarm optimization has been used for evolving optimized portfolio asset allocations in a volatile market condition. The proposed approach is centered on optimizing the Conditional Value-at-Risk (*CVaR*) (Rockafellar and Uryasev, 2002) measure in different market conditions based on several objectives and constraints. Other than implementing the general definition of *CVaR* and its minimization formulas associated, the authors have concentrated here into dealing with fully discrete distributions enhancing the usefulness and properties of *CVaR* in case furnishing the elementary way of calculating *CVaR* directly. The results are compared with those obtained with the optimiza-tion of Value-at-Risk (*VaR*) measure of the portfolios under consideration. A comparative application of the proposed approach along with the *VaR* approach is demonstrated on a collection of several financial instruments enabling with a distributional assumption for employing the particular type of financial assets for developing a much more generalized framework.

The authors have planned the article into the following different sections providing an overview of the conventional concept of Conditional Value-at-Risk in section 2. Section 3 then elucidates the mathematical formulation of the *VaR* and the *CVaR* measure. A discussion of the particle swarm optimization procedure along with the algorithm associated is provided in section 4. The findings of the work are summarized in section 5. Conclusions added to future directions of research are drawn out finally in section 6.

2. CONVENTIONAL CONCEPT OF CONDITIONAL VALUE-AT-RISK

Introduction of return risk management framework by Markowitz (1952) (Markowitz 1952) has come a long way in the process of portfolio optimization. Of late, the usage of the alternative coherent technique is been done for the reduction of the probability in incurring large amount of losses by a portfolio. This can be done by the assessment of the specific loss that will be exceeding the value at risk. The outcome risk measure is termed to be the Conditional Value-at-Risk (*CVaR*) (Rockafellar and Uryasev, 2002), thanks to the evolving fields of data intelligent management and archival techniques in industrial portfolio management. Simulation by two basic requirements in turn has developed in the portfolio optimization procedure by (i) risks, constraints and adequate modeling of utility functions and (ii) efficient handling of huge numbers of scenarios and instruments. In terms of mathematics, derivation of *CVaR* is done by considering the values of the weighted average at the intervals of the value-at-risks and the losses exceeding the value-at-risks. Being *CVaR* compared to *VaR*, it not only traces several different loss distributions and can even also be easily expressed in minimization formula.

Measures of risk plays a vital role notably in grappling with losses which might have been incurring in finance under the shed of uncertain conditions. Loss, being derived as a function $z = f(x, y)$ of a decision vector $x \in X$ purporting to depict the values of a number of variable viz. interest rates or weather data in terms of the future values. If y is assumed to be random with accepted probability distribution z, then it produces to be as a random variable having its dependent distribution on the superior of x. If any optimization problem shows the involvement of z in turn of the superior of x, then it can be accounted not as just expectations but even as "riskless" of x.

Percentile measures of loss or reward can be done by $f(x,y)$, which is taken to be the loss function rely upon the decision vector $x = (x_1, ...x_n)$ and the random vector $y = (y_1, ...y_m)$, then *VaR* can be calculated as α-percentile, representing the loss distribution which is considered to be the smallest value where the probability which loss exceeds or is equal to the value which is greater or equal to α. In such case $CVaR^+$ which is also known to be the "upper *CVaR*" is the expected loss which strictly exceeds *VaR* (in turn known as Mean Excess Loss) and Expected Shortfall. *CVaR*-in turn known as "lower *CVaR*" is the expected loss weekly exceeding *VaR*, which is the expected loss equal to or exceeding *VaR*. It is also known as Tail *VaR*.

Thus, *CVaR* is the weighted average of *VaR* and *CVaR*⁺ (Rockafellar and Uryasev, 2002). It can be derived from the following formulation:

$$CVaR = \lambda VaR + (1 - \lambda) \, CVaR^+ , \; 0 \le \lambda \le 1 \qquad \qquad \ldots (1)$$

where, λ is the Lagrange multiplier.

Since *CVaR* is convex as shown in Figure 1, *VaR*, *CVaR*⁺, *CVaR*⁻ can also be non-convex. This shows credible inequalities as:

$$VaR \le CVaR^- \le CVaR \le CVaR^+ \text{ (Rockafellar and Uryasev, 2002)}$$

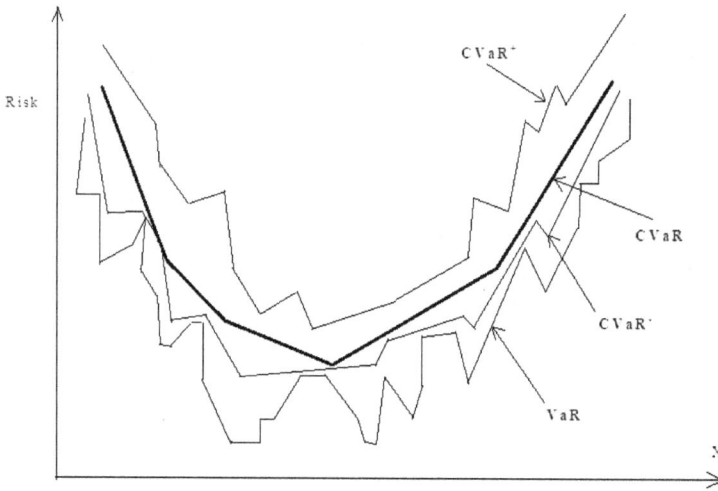

Fig. 1: CVaR: Convex Function (Rockafellar and Uryasev, 2002)

The relationships between *VaR*, *CVaR*, *CVaR*⁻ and *CVaR*⁺ are shown in Figure 2.

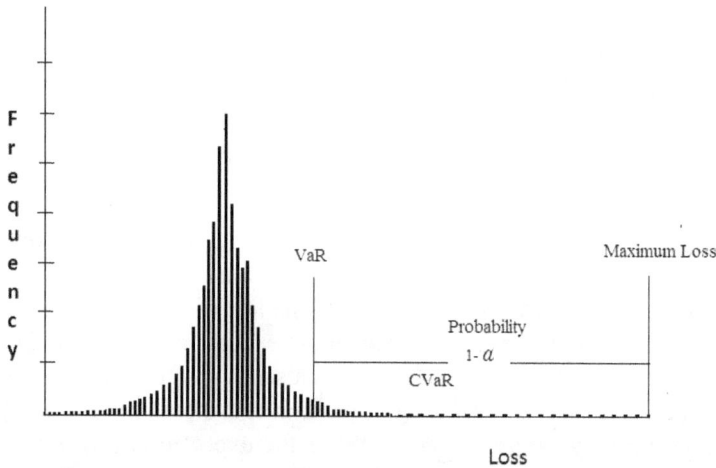

Fig. 2: *VaR, CVaR, CVaR⁺, CVaR⁻* (Rockafellar and Uryasev, 2002)

3. MATHEMATICAL FORMULATION

VaR is considered to bean important measure for the disclosure of a stipulated financial portfolio in terms of varied risk situations, instinctive in financial structures which in turn are also considered to be of paramount importance in the portfolio optimization purposes.

Considering a portfolio *P*, levelheaded by *k* assets, $S = \{S_1, S_2,..., S_k\}$, and $W = \{W_1, W_2...W_k\}$ considered as the relative weights or portions of the assets in the stipulated portfolio, hence the price can be computed as:

$$P(t) = \sum_{i=1}^{k} S_i(t)W_i$$

... (2)

In which $S_i(t)$ and W_i stand to be the values and importance levels of the portfolio at a given time period *t*, accordingly.

The *VaR* of the portfolio *P*, which is the utmost wonted loss over an extent of time at a given level of confidence (α), then can be considered to be the minimal number l such that the probability that the loss *L* surpass *l* is not greater than $(1 - \alpha)$, i.e.

$$VaR_\alpha = \inf\{l \in R : P(L > l) \le 1 - \alpha\} = \inf\{l \in R : F_L(l) \ge \alpha\}$$

... (3)

A plethora of techniques and models for estimation of *VaR* from the time horizon, level of confidence and the unit of *VaR* is available in the literature (Holton, 2003; Jorion and Philippe, 2001; Pearson, 2002; Glasserman, 2004; Rouvinez, 1997; Wilsonand Alexander, 1999).

All the techniques and models depend on a set of assumptions of their own. However, the most common assumption stands to be the best estimator for future changes in market conditions is the historical trace of available market data. Some of the well-known models for estimating *VaR* include:

1. *Variance-Covariance (VCV) Model*—It is helpful for the assumption of the risk factor returns which is to be normally (jointly) distributed in every cases, and at the same time the portfolio return which in turn also to be normally distributed. It is also helpful in assumption of the modification in the worth of portfolio which is directly contingent on all risk factor returns. In the beginning of 1990s, J.P. Morgan popularized the variance-covariance or the delta-normal model. The assumption of the portfolio return to be normally distributed gives an indication of composition of assets in the portfolio. The changes/deltas being linear state the change in the portfolio value which in turn is directly defenseless on all the alteration in the values of the assets. This implies that the portfolio return is also linearly dependent on all the asset returns and that the return on assets further jointly been normally distributed. With further assumption of the only risk factor associated with a stated financial portfolio is the value of the portfolio itself, the 95% confidence level *VaR* for *N* assets over a holding period, is given by

$$VaR = -V_p(\mu_p - 1.645\sigma_p)$$

... (4)

In which, the mean μ_p is given as,

$$\mu_p = \sum_{i=1}^{N} \varpi_i \mu_i$$

... (5)

The standard deviation σ_p is given as,

$$\sigma_p = \sqrt{\Omega^T \Sigma \Omega} \qquad \qquad \dots (6a)$$

$$\Omega = \begin{bmatrix} \varpi_1 \\ \varpi_2 \\ \varpi_3 \\ \cdot \\ \cdot \\ \cdot \\ \varpi_N \end{bmatrix} \qquad \qquad \dots (6b)$$

$$\Omega^T = \begin{bmatrix} \varpi_1 & \varpi_2 & \varpi_3 & \cdot & \cdot & \varpi_N \end{bmatrix} \qquad \qquad \dots (6c)$$

where, i refers to the return on asset i and p refers to the earnings on the portfolio for standard deviation (σ_p) and mean (μ_p). V_p is taken to be the value of portfolio at the beginning (in currency units). ϖ_i is hence considered to be the ratio of V_i and V_p.

If compact and maintainable data set has been purchased from third parties, VCV model stands to be beneficial in terms of their usage and also in the speed of calculation by the usage of optimized linear algebraic libraries. The main drawbacks of this model lie in the assumption that the portfolios generally comprise assets whose delta are linear and that the market price returns/asset returns are normally distributed.

2. *Historical Simulation (HistSim) Model*—Being emerged as the industry standard for computing *VaR*, it is established on the assumption that the expected returns on asset will always bear an equal amount of distribution as in the past. Hence, *HistSim* is considered to be the simplest and most transparent method for calculating the *VaR*. This model for computing a percentile (*VaR*), involves operating the current set of portfolio cross wide of a set of historical trace for yielding modifications in the portfolio value. Its simplicity of implementation stands to be its most important benefit along with not assuming a normal distribution of asset returns like the VCV model. Its intensive calculation computationally along with the requirement for a large market database fall under its main drawbacks. In *HistSim*, *VaR* is evaluated as:

$$VaR = 2.33 M \sigma_p \sqrt{10} \qquad \qquad \dots (7)$$

In which, M is considered as the market value of the portfolio and σ_p is considered to be the historical volatility of the portfolio. The constant 2.33 stands for the number of σ_p which is required for a level of certainty of 99% and the constant $\sqrt{10}$ refers to the number of days in the holding period.

Basically, computation of VaR is done in the *HistSim* method in two simple steps. Firstly, construction of a series of pseudo-historical portfolio returns is been calculated using today's portfolio weights and historical asset returns. Secondly, the computation of the *VaR* and the current asset returns quantile of the pseudo-historical portfolio returns are carried out.

3. *Monte Carlo Simulation*—This model basically undergoes the random simulation of future asset returns. Usage of this simulation is done generally for the computation of *VaR* for portfolios which are holding the securities with non-linear returns and in which the computational effort required is non-trivial. Conceptually, simplicity of this method stands to be its added advantage, but rather it is computationally more intensive than both the VCV and *HistSim* models. The generic Monte Carlo *VaR* calculation incorporates the following steps.

   (a) Predefining N, denoting the number of iterations which is to be performed.
   (b) In consideration of every iteration in N,
     • Generating a random sequence of events of market which moves by the usage of some existing model present in the market.
     • Revaluing the portfolio covered by theassumed market volatility sequence of events.

   (c) Computing the portfolio profit or loss (PnL) in case of the assumed sequence of events and for doing so, subtracting the ongoing market value of the portfolio from its market value which has already been computed in the last previous steps.
   (d) Sorting the result *PnLs* required for obtaining the simulated Profit and Loss (*PnL*) distribution for the portfolio.
   (e) Finally, calculation of *VaR* at a particular level of confidence with the usage of the function of percentile.

The features of *CVaR* represent the risks which are simple and convenient in nature hence measuring the downside risks, and are applicable to non symmetric distribution of losses. Stable statistical estimates of *CVaR* appear to be its integral characteristics in comparison to *VaR* which can get influenced by any scenario. *CVaR* yields values in a continuous process in terms of confidence level α, steady at different levels of confidence in comparison with *VaR* (*VaR*, *CVaR⁻*, *CVaR⁺* may not be continuous to α). *CVaR* portfolios coincide in case of normal distribution of loss in optimal variance to the level of consistency in mean variance approach. *CVaR* is variedly acceptable due to its easy control and optimization process for non normal distributions, even shaping of loss distribution is being done using *CVaR* constraints for the first online procedures.

4. PARTICLE SWARM OPTIMIZATION

The occurrence of evolutionary computation has been inspiring fresh assests for optimization in different problem solving procedures variably in the field of portfolio management. Evolution of algorithms, such as Genetic Algorithm (GA) (Goldberg,1989), Ant Colony Optimization (ACO) (Dorigo *et al.,* 1996), Simulated Annealing (SA) (Jiang *et al.,* 2007) and Particle Swarm Optimization (PSO) (Clercand Kennedy, 2002; Krusienski and Jenkins, 2005; Kennedy and Eberhart, 1995; Robinson and Yahya, 2004; Kennedy *et al.,* 2001), all of which tend to find the global solution of a stated problem.These algorithms have been used as effective tools for evaluation of numerous points in the search space simultaneously.

Established on the simulation of simplified social models *viz.* bird flocking, fish schooling, and the swarming theory, PSO stands to be evolutionary computation mastery in terms of individual enhancement along with population cooperation and competition. The concept craves only primeval mathematical drivers, which is not at all computationally expensive in consideration to memory necessities and irrespective of time. Optimization of fitness function is been evaluated for every particle. The desired value for the comparison purpose of the particle's fitness worth with particle's *pbest* (personal best) is found out and if current worth is excelling than *pbest*, then *pbest* valueis then been set which is balanced to the on going worth and the *pbest* position equal to the current position in a *K*-dimensional space. Comparing particle's fitness worth with the particle's fitness values obtained so far falls into the next procedure. If the ongoing worth is better than the *gbest* (global best), then the *gbest* value is being reset to the current particle's value. Repetition of this process continues just before a user-defined staying criterion is been arrived at.

In the process of PSO, the representation of each potential particle representing as a particle with a position vector **x**, refers to the phase weighting factor **b** and a moving velocity **v**, respectively. For *K* dimensional optimization, the position and velocity of the ith particle has been depicted as $\mathbf{b}_i = (b_{i,1}, b_{i,2}, ..., b_{i,K})$ and $\mathbf{v}_i = (v_{i,1}, v_{i,2}, ..., v_{i,K})$, respectively. Each particle having its own best position $b_i^P = (b_{i,1}, b_{i,2}, \cdots, b_{i,K})$ which corresponds to the distnctive best objective value which is obtained till the time *t*,is referred to as *pbest*. The global best (*gbest*) particle is expressed by $\mathbf{b}^G = (b_{g,1}, b_{g,2}, \cdots, b_{g,K})$, which is represented as the best particle so far at time *t* in the complete swarm. The fresh velocity $\mathbf{v}_i(t+1)$ for particle *i* is been reconditioned by,

$$v_1(t + 1) = wv_i(t) + c_1[b_i^P(t)] + c_2[b^G(t) - b_i(t)] \qquad\qquad \text{... (8)}$$

where, *w* is considered as *inertia weight*, $v_i(t)$ is considered to be the traditional velocity of the particle *i* at time period *t*. Seemingly from the above sated equation, the newly obtained velocity is connected to the velocity which is obtained in the previous course which in turn is denoted by *w* (weight) and is also related to the position of the particle with that of the global best one by acceleration constants c_1 and c_2. Accelerations c_1 and c_2, being the constants adjust to the amount of tension in PSO system. High values resulting in unexpected movement towards the target regions along with low values allow the particles to roam far from the target regions before being tugged back. The acceleration constants c_1 and c_2 are therefore been cited as the cognitive and social rates for representing as the weighting factor of the acceleration terms, pulling the individual particle towards the personal best and global best positions. Particle swarm optimization is thus a population-based stochastic optimization procedure which has been influenced by social behavior of bird flocking or fish schooling. Thus PSO is considered as each single solution like a "bird" (particle) in the search space of food (the best solution). All particles having fitness values are appraised by the fitness function, and have velocities that direct the "flying" (or evaluation) of the particles which in turn gets boot up with a set of random particles (solutions). PSO peruse to find the optimal resolution by amending generations in every iteration. All particles are re-established by two "best" values already entrenched as the *pbest* or personal best connoting the best solution or fitness which a particle has resolved so far while the other one is the *gbest* or global best demonstrating the best value achieved by any particle in the population. The best value achieved in the topological neighbor or on part of a population is a local best and is called pbest. Finding the two best values, the velocity and positions of the particle are updated using following equation (Bhattacharyya and Maulik, 2013).

$$V_{t+1}iX_{(t+1)l} = wV_{ti} + c_1 \text{rand}_1()(pbest_i - X_{ti}) + c_2 \text{rand}_2()(gbes_{ti} - X_{ti}) = X_{ti} + V_{(t+1)l} \quad \dots (9)$$

Where, i is once again contemplated to be the index of each particle, t is the prevailing iteration number, $\text{rand}_1()$ and $\text{rand}_2()$ are random numbers between 0 and 1. $pbest_i$ is the best former occurrence of the ith particle while $gbest_i$ is the best particle between the whole population. Constants c_1 and c_2 are the weightage factors of the stochastic acceleration terms, which in turn pull each particle toward the $pbest_i$ and $gbest_i$, w being the inertia weight governing the exploration properties of the algorithm. If $c_1 > c_2$, the particle tends to reach $pbest_i$, the best position labeled by the particle, rather than converge to $gbest_i$ established by the population and vice versa. The flow diagram of the PSO algorithm is shown in Figure 3.

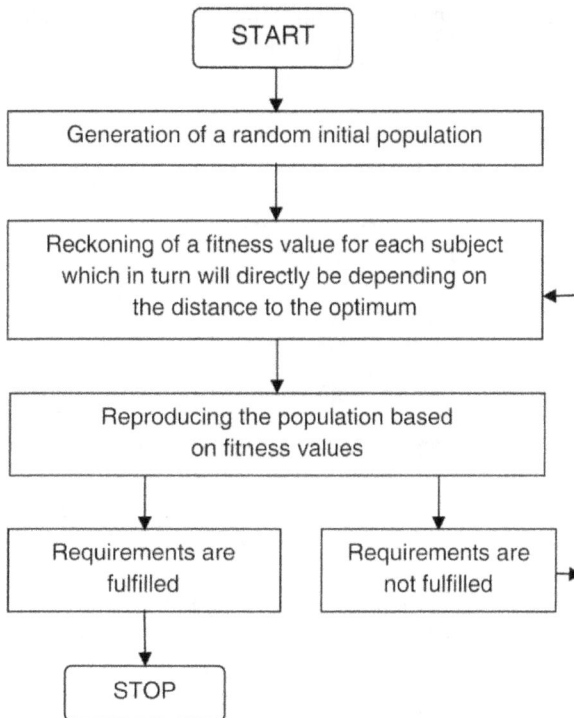

Fig. 3: Flow Diagram of PSO Algorithm

The above procedure makes us learn that PSO starts with a class of a population bring about randomly, both in turn having fitness values for evaluating the population. The population can be updated and searched for the best with random techniques in both the procedures without any success been guaranteed.

However, unlike the genetic algorithm, PSO does not include genetic operators such as crossover and mutation. With the internal velocity, particles start updating themselves having memory, which stands to be of utmost importance to the algorithm. PSO proves to be very different significantly, when terms of comparison arises to genetic algorithms (GAs). Chromosomes in turn part facts or news with each other for moving whole population like one group enhancing an optimal area where only *gbest* (or *lbest*) gives out the information to others. Hence it proves to be a one-way information sharing mechanism in turn proving the evolution just looking for the best solution, along with all the particles tending for coverage to the best solution quickly even in the local version in most of the cases.

5. PROPOSED METHODOLOGY

The proposed approach is centered around the optimization of the Conditional Value-at-Risk (*CVaR*) measures of a portfolio comprising several financial instruments at different market conditions based on several objectives and constraints. Application of PSO for *CVaR* optimization is demonstrated with reference to the minimization of the risks involved in the portfolios under consideration, thereby minimizing the portfolio losses incurred. The flow diagram of the proposed methodology is shown in Figure 4.

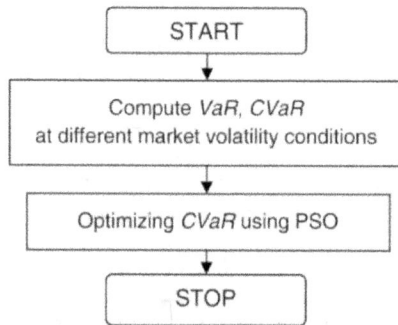

Fig. 4: Flow Diagram Depicting the Proposed Methodology

The procedure of portfolio asset allocation optimization is demonstrated on a collection of 20 portfolios with several asset variations. Here in this process of optimization, the particle swarm optimization algorithm has been run with two different numbers of generations *viz.,* 500 and 1000 with the constants already been specified in Table 1.

Table 1: Particle Swarm Optimization Parameters Employed

Sl. No.	PSO Parameter	Values Used
1.	Number of Generations	(500, 1000)
2.	Inertia Weight	0.8
3.	Acceleration Coeffient ($\phi1$)	1.5
4.	Acceleration Coeffient ($\phi2$)	1.5

The optimization of the portfolio asset allocation is achieved with particle swarm optimization (Ying et *al.,* 2006; Christopher, 2001). In order to faithfully allocate assets in a given level of confidence, it minimizes the Conditional Value-at-Risk (*CVaR*) of the portfolio using the following function as the fitness function.

$$CVar = \frac{e^{-(\frac{VaR^2}{2})}}{a\sqrt{2\pi}} \qquad \ldots (10)$$

where, $a = 0.01$ considering a confidence level of 99% and VaR are the Value-at-Risk measures of the portfolios under consideration.

Table 2 lists the different archived average optimized portfolios over two different numbers of generations along with their costs for a confidence level of 99%.

In addition, as a comparative study, the Historical Simulation (*HistSim*) model has been used for computing VaR of the portfolios under consideration. The particle swarm optimization algorithm is then used to obtain optimum VaR measures of the portfolios using equation (7) as the fitness function. The optimized VaR values obtained using the particle swarm optimization algorithm are also listed in Table 2 for the sake of comparison.

Table 2: Comparative Results of Optimized Portfolios with Their Costs, CVaRs and VaRs at a Confidence Level of 99%

Portfolio No.	Portfolio Cost (in currency units)	CVaR	VaR
1.	26758.782629	0.190600	0.224732
2.	29942.989518	0.185829	0.226188
3.	34868.339288	0.183854	0.224514
4.	36591.431339	0.184595	0.224596
5.	35996.607129	0.185289	0.224314
6.	32243.667175	0.181077	0.224495
7.	30576.016315	0.182792	0.225213
8.	33300.584075	0.186494	0.225019
9.	30446.646374	0.181387	0.225806
10.	29874.663429	0.187987	0.225377
11.	36049.580870	0.182420	0.224926
12.	33303.676941	0.184245	0.225910
13.	38518.502646	0.179264	0.224766
14.	47675.001370	0.177929	0.224173
15.	26358.703336	0.193556	0.226707
16.	30098.635906	0.189505	0.225343
17.	35391.282927	0.184244	0.224860
18.	24874.327332	0.182948	0.224615
19.	33692.352544	0.184193	0.225842
20.	36283.390566	0.180552	0.224320

From Table 2, it is evident that the *CVaR* measures provides a more realistic impression regarding the allocation of financial instruments since for all the 20 portfolios, the *CVaR* measures reflect minimized market risks as compared to their *VaR* measures.

It can be observed that PSO is able to provide a faithful selection/allocation strategy of assets in the financial portfolio which can be derived by the usage of an optimization procedure in particular iteration and generated a solution which is local optimum. Moreover, PSO further delve into to find much optimized solution using *CVaR* techniques than the one spwaned by PSO by usage of *VaR* techniques. The proposed approach aims at minimizing the *CVaR* measures of the portfolios under consideration using particle swarm optimization. The optimized portfolios are seen to outperform the corresponding *VaR* based optimized portfolios as regards to minimization of market risks. To the best of our knowledge, no such attempts have been reported in the literature so far. Hence, this initiative is a maiden venture in this direction.

6. CONCLUSION

In the fields of economics and finance, portfolio management has been assumed to be of supreme importance as a systematic discipline given at the strategies for diversification in investment. This article attempts to evolve a selection and allocation strategy of portfolios in a volatile market condition by means of the optimization of the Conditional Value-at-Risk (*CVaR*) measures of the portfolios under consideration. A particle swarm optimization procedure is adopted on historical portfolio data for this purpose. Faithful selection of results is exhibited on a collection of 20 different portfolios with several asset combinations.

CVaR being into a new risk measuring procedure provides significant advantages when been compared to *VaR* which in turn is able to quantify risks beyond the level of *VaR*. Hence it is known to be a coherent risk measurement procedure which is consistent in different confidence levels. The authors have attempted of evolving a strategy for selection and allocation of portfolios under different financial market conditions. For the stated purpose Particle Swarm Optimization (PSO) procedure is adopted under the above two mentioned conditions .Thus the proposed approach is targeted at minimizing the *CVaR* measures of the portfolio under consideration.

Methods however remain to be investigated to incorporate the aspect of return maximization in the portfolio allocation scenario through multi-objective optimization techniques. The authors are currently engaged in this direction.

REFERENCES

Bai, D., Carpenter, T. and Mulvey, J. (1997). Making a case for robust optimization. *Management Science* (43) pp. 895–907.

Bhattacharyya, S. and Maulik, U. (2013). *Soft Computing for Image and Multimedia Data Processing.* Springer Verlag Heidelberg, Germany.

Birge, J.R. and Rosa, C.H. (1995). Modeling investment uncertainty in the costs of global CO_2 emission policy. *European Journal of Operational Research* (83) pp. 466–488.

Brown, A. (2004). "The Unbearable Lightness of Cross-Market Risk," *Wilmott Magazine.*

Cariño, D.R. and Ziemba, W.T. (1998). Formulation of the Russell-Yasuda Kasai financial planning model. *Operations Research* (46) pp. 433–449.

Christopher, C. (2001). *Time Series Forecasting,* Chapman and Hall/CRC, pp. 181–214.

Clerc, M. and Kennedy, J. (2002). "The particle swarm-explosion, stability and convergence in a multi-dimensional complex space," *IEEE Transactions on Evolutionary Computation* (6) (1) pp. 58–73.

Dorigo, M., Maniezzo, V. and Colorni, A. (1996). "Ant System: Optimization by a Colony of Cooperating Agents," *IEEE Transactions on Systems, Man, and Cybernetics–Part B.* (26) (1) pp. 29–41.

Dowd, K. (2005). *Measuring Market Risk.* 2nd edition, John Wiley and Sons.

Glasserman, P. (2004). *Monte Carlo Methods in Financial Engineering.* Springer.

Goldberg, D. (1989). *Genetic Algorithms in Search, Optimization and Machine Learning.* MA: Addison-Wesley Professional.

Hiller, R.S. and Eckstein, J. (1993). Stochastic dedication: Designing fixed income portfolios using massively parallel benders decomposition. *Management Science* (39) pp. 1422–1438.

Holton, G.A. (2003). *Value-at-Risk: Theory and Practice.* Academic Press.

Jorion, P. (2001). *Value at Risk.* 2nd edition, McGraw-Hill.

Jorion and Philippe (2001). *Value at Risk: The New Benchmark for Managing Financial Risk.* 2nd edition, McGraw-Hill Trade.

Kennedy, J. and Eberhart, R.C. (1995). "*Particle swarm optimization,*"Proc. IEEE Neural Networks IV. Pisctway, NJ, pp.1942–1948.

Kennedy, J., Eberhart, R.C. and Shi, Y. (2001). *Swarm Intelligence.*San Mateo, CA: Morgan Kaufmann.

Krusienski, D.J. and Jenkins, W.K. (2005). "*Design and performance of adaptive systems based on structured stochastic optimization strategies,*" *IEEE Circuits and System Magazine,* (5) pp. 8–20.

Markowitz, H.M. (1952). "Portfolio selection," *The Journal of Finance,* (7) (1) pp. 77–91.

McNeil, A., Frey, R. and Embrechts, P. (2005). *Quantitative Risk Management: Concepts, Techniques and Tools.* Princeton University Press.

Mulvey, J.M., Vanderbei, R.J. and Zenios, S.A. (1995). Robust optimization of large-scale systems. *Operations Research* (43) pp. 264–281.

Pearson, N.D. (2002). *Risk Budgeting.* John Wiley and Sons.

Robinson, J. and Yahya, R.S. (2004). "Particle swarm optimization in electromagnetic," *IEEE Trans. Antennas and Propagation* (52) pp. 397–407.

Rouvinez, C. (1997). "Going Greek with VAR," *Risk,* (10) (2) pp. 57–65.

Rockafellar, R.T. and Uryasev, S. (2002). *Journal of Banking and Finance* (6) pp.1443–1471.

Stambaugh, F. (1996). Risk and value at risk. *European Management Journal.* (14) pp. 612–621.

Vladimirou, H. and Zenios, S.A. (1997). Stochastic linear programs with restricted recourse. *European Journal of Operational Research* (101) pp. 177–192.

Wilson, T. and Alexander, C. Ed. (1999). "Value at risk" in *Risk Management and Analysis.* Chichester, England, Wiley (1) pp. 61–124.

Ying, F., Yi-ming, W. and Shang-jun, Y. (2006). *Complexity in Financial System: Model and Analysis.* Beijing: Science Press.

A Study of Factors Influencing Adoption of ISO 14000 Certification in MSMEs in Coimbatore

V.D. Krishnaveni[1] and R. Nandagopal[2]

PSG Institute of Management, PSG College of Technology,
Coimbatore–641 004, Tamil Nadu
E-mail: [1]krishnaveni.damodaran@gmail.com; [2]nandagopal@psgim.ac.in

ABSTRACT: *The ISO 14000 refers to international standards that deal with the establishment and implementation of Environmental Management Systems in organizations. Coimbatore is an industrial hub with more than 25,000 small, medium and large scale enterprises. This study aims to find the prevalent perceptions or opinions about ISO 14000 standards in the manufacturing sector, specifically engineering ancillary, in Coimbatore and also to measure the intention of firms to adopt the same.*

The research design was exploratory in nature. Convenience sampling was used and the sampling frame was restricted to engineering ancilliary units with less than 75 employees. Data collection involved a survey using a questionnaire. The statistical tools employed were Karlpearson's Correlation, Regression Analysis and t-tests using SPSS software.

Statistical analysis helped to establish that the influence of buyer pressure, organisation attitude and difficulty in certification on intention to adopt ISO 14000 levels is either weak or nonexistent. The analysis of the data by the respondents in the manufacturing sector in Coimbatore suggests that there are low levels of adoption of ISO 14000 standards as it is not accepted as an indicator of environmental performance by several micro and small enterprises.

Keywords: ISO 14000, MSME, Buyer Pressure, Organization Attitude, Difficulty in Certification.

1. INTRODUCTION

The ISO 14000 family of standards represents standards used by firms and other organizations in designing and implementing an effective Environmental Management Systems. This paper attempts to study the factors that help or hinder the adoption of these standards in the micro and small scale enterprises in the Coimbatore region.

1.1 International Organisation for Standardisation

Post World War II, the International Organisation for Standardisation (ISO) was set up under the aegis of the newly formed United Nations Standards Coordinating Committee to replace the earlier existing International Federation of National Standardising Associations (ISA). Since its inception in 1947, the ISO has been working towards promoting worldwide industrial and commercial standards. As of today, it has a membership of 165 member countries.

The standards are world-class specifications for products, services and systems, to ensure quality, safety and efficiency. According to the ISO, they refer to documents that "provide requirements, specifications, guidelines or characteristics that can be used consistently to ensure that materials, products, processes and services are fit for their purpose" (International Standards Organisation). Such global standards covering a wide range of industries from manufacturing to healthcare aid in facilitating international trade.

ISO 9000

The ISO 9000 refers to a family of standards that deal with quality management systems. It is based on eight management principles and enables organizations to ensure that customer and stakeholder needs are met along with statutory and regulatory requirements in a product. The ISO 9001 standard is the most popular among all ISO standards and was implemented in over 1 million organizations by December, 2009. The certification helps firms gain customer confidence and hence a growth in client base.

ISO 14000

The ISO 14000 refers to a family of international standards that were released in 1996 to look into the environmental aspects of operations and product standards. It targets at controlling environmental impact and improving environmental performance. Before its development, organizations voluntarily developed their own Environmental Management Systems (EMS), but such systems were hard to compare and so the ISO 14000 standards were created to facilitate international trade.

The ISO 14001 standard provides guidelines for the establishment or improvement of an Environmental Management System and can be certified. The benefits of implementing the ISO 14001 include (ISO 14000 - Environment Management):

- Reduced cost of waste management
- Savings in consumption of energy and materials
- Lower distribution costs
- Improved corporate image among regulators, customers and the public.

However, ISO 14000 standards have their share of critics who state that they are de facto trade barrier as environmental improvement and truly "free" trade are incompatible ideals (Murray, 1997). Even though the ISO 14000 standards are controversial, the ISO process is becoming more participatory with every member country having an opportunity to be represented in negotiations (Haufler, n.d.).

World over businesses are beginning to realize that they cannot relegate environmental protection to a few CSR activities. The need for environmental concern to be integrated into the procurement, production and distribution systems has been realized and firms have taken to it in a big way. Today, governments are talking about Extended Producer Responsibility where organizations have to take care of used products discarded by customers and find ways of disposing/recycling them in an environmentally sustainable manner.

Given the high levels of environmental awareness among customers, there is an increasing amount of pressure on the suppliers to conform to environmental standards. This study is significant as it analyses the intent of the medium and small scale manufacturing enterprises in Coimbatore to adopt the ISO 14001 standards which would be an expression of the firms' commitment to reduction of adverse environmental impacts.

2. REVIEW OF LITERATURE

ISO Standards were created with the intention of facilitating international trade and commerce. While the ISO 9000 standards dealing with quality management systems became extremely popular, the ISO 14000 is yet to catch up in terms of number of firms that are certified.

2.1 ISO 14000

Globally, there has been an increase in the awareness of adverse environmental impact due to industry practices. Governments and law makers are trying to curb this by creating statutory and increasingly stringent laws. Organisations have however, also decided to contribute their mite by voluntarily adopting certain international standards set by third party organizations in this regard. Once viewed as an external cost, it has now moved on to become a source of competitive advantage to firms (Ratiu and Maria, 2014).

The British Standards Institute defined an EMS as "the organisational structure, responsibilities, practices, procedures, procedures, processes and resources for determining and implementing environmental policy" (Marett, 2000). It is essentially a management system which tracks the organisation's environmental objectives and ensures that negative impacts are reduced to acceptable limits. As many different organizations were establishing their own EMSs, it was becomingly increasingly difficult to compare the various systems leading to the creation of the ISO 14000 standards.

The evolution of the ISO 14000 standard has been summarized in the figure below (Arvanitoyannis and Boudouropoulos, 1999):

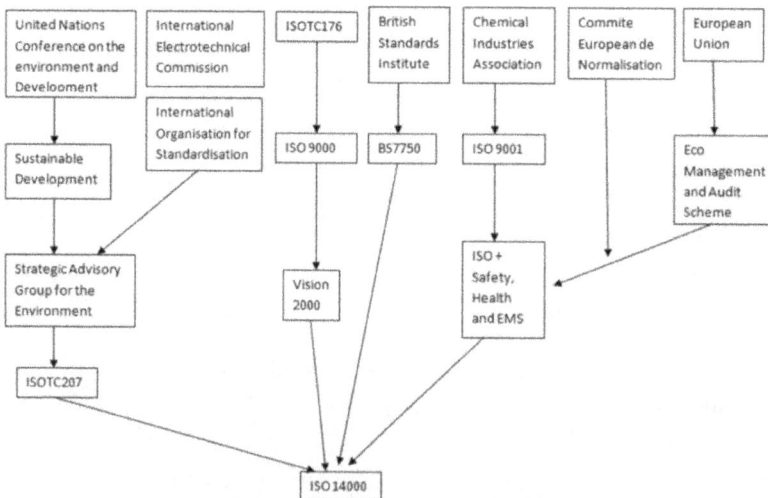

Fig. 1: Evolution of the ISO 14000 Standards
(adapted from Arvanitoyannis and Boudouropoulos, 1999)

At the time of creation, the standard catered to three aspects: evaluation and auditing tools, management system standards and product oriented support tools (Marett, 2000).

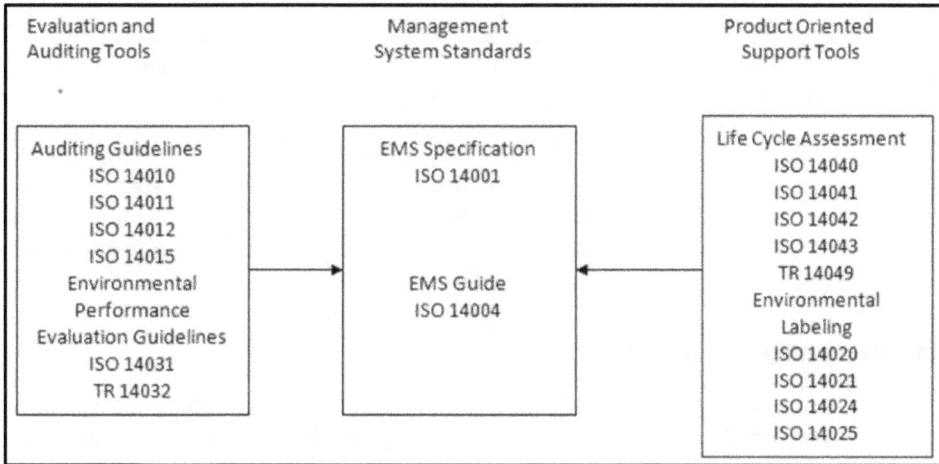

Fig. 2: The Structure of the ISO 14000 Series of Standards (adapted from Marett, 2000)

The ISO 14000 EMS Model is based on Deming's Plan-Do-Check-Act model as depicted below (Pawliczek and Piszczur, 2013):

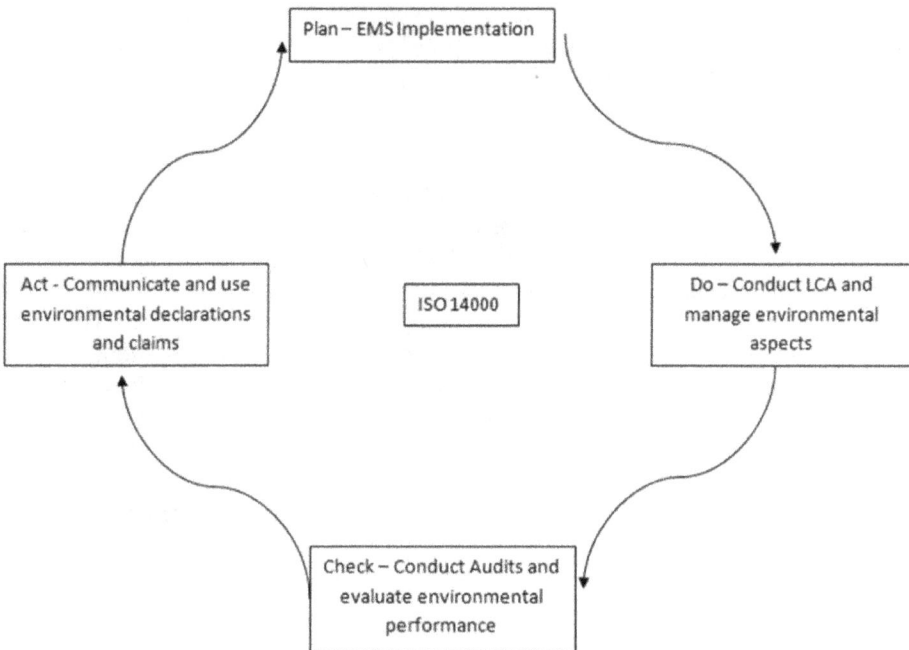

Fig. 3: Environmental Management System Model by ISO 14000 (adapted from Pawliczek and Piszczur, 2013)

An organisation has the option of self-declaring that it meets the requirements of the standard. Usually, however, organizations pass a third party audit of their EMS to become registered to ISO 14001.

While proponents of ISO 14000 believe that it is pro-active, they need to realize that the standards are only active in reducing the impact on the environment and therefore in a sense, they are reactive to the global environmental destabilization scenario (Ball, 2002).

The ISO 9000 and ISO 14000 standards followed effective diffusion patterns and have been implemented in several countries around the world. Slowly, however, there seems to be a decertification trend setting in. While the ISO states the reason for this as failure to conduct recertification audit, further research has to be done to understand the failure on the part of the firms (Marimon, Heras, and Casadesús, 2009).

2.2 Drivers of ISO 14000 Certification

Even before the introduction of the ISO 14000 standards it was seen as an image builder, a means of reducing product and process costs, lifting the regulatory burden and avoiding liability (Pouliot, 1996).

A comparative study of firms in Japan, Taiwan, Hong Kong and South Korea showed that the drivers for ISO 14000 certification were cost reductions, increased productivity, quality improvements, environmental improvements, increased on time delivery to customers, increased customer satisfaction, increased market share, increased profits, improved internal procedures, improved employee morale, improved relations with authorities, improved relations with communities and improved corporate image (Pan, 2003). Other drivers are costs and uncertainties involved at the firm level, improving risk management and lowering liabilities and harmonizing standards with ISO 9000 (Vastag, 2003).

Different forces were affecting the adoption of the standards in different countries. In Brazil, the dominant factor was the complexity of EMSs for small and medium firms. In China, pressure from foreign buyers was a compelling reason (Hariz and Bahmed, 2013). Critical factors for the successful implementation of an EMS are planning, strategy, capability building and process management. Potential barriers identified among New Zealand firms were implementation processes and costs, external engagement, information, infrastructure and contractor commitments (Cassells, Lewis, and Findlater, 2012). A study among managers in Saudi Arabia states that the five driving forces are employee morale, efficiency and quality, relations with government, safety of products and relations with environmental organizations. Difficulties associated are related to costs such as fees of consultants, costs of changing within organizations, costs of maintaining the system, fees of certification agencies and costs of internal auditing (Kadasah, 2013). Increased operation and business performance of firms was perceived as an advantage of adopting ISO 14000 certification among Malaysian firms (Wahid and Y, 2010).

A colloquium in Canada suggested that the practical challenges of implementing ISO 14001 revolve around critical loops and synergies among the management system elements, management system auditing and integrating management systems (Searcy *et al.*, 2012). One study has shown that voluntary protection programmes resulted in workplace injury and illness rates that were below industry standard (Cascio and Baughn, 2000). An American study suggests that past experience with Total Quality Management, past experience with ISO 9000 and QS 9000 and the current status of cross-functional programmes influences the decision to undergo ISO 14001 certification (Curkovic, Sroufe, Melnyk, and Montabon, 2001).

ISO 14001 standard does not specify any performance goals and the language may be interpreted ambiguously, complicating implementation. This has been perceived to be a weakness (Elefsiniotis and Wareham, 2005). Also in the case of SMEs, which do not operate under very formal systems, ISO 14000 is seen as simply increased paperwork and costs. Many SMEs do not have the knowledge or the resources to implement EMSs and market incentives are insufficient (Susan, 2007). In fact, it is frequently suggested that larger firms are frequent adopters of EMS (Cassells, Findlater, and Lewis, 2009). In England and Ireland, implementing EMSs improved onsite environmental management but did not translate into better environmental performance ("ISO Guidance for Small and Medium-Sized Businesses", 2006). Certification timeframe was an important barrier to implementation according to a study in the United Kingdom (Strachan, Sinclair, and Lal, 2003).

2.3 Misconceptions about ISO 14000

Certain misconceptions exist about the ISO and the ISO 14000 standards in particular. The ISO is not the organisation that issues a certificate acknowledging compliance with its standards. This is done by third party organizations which undergo a certification process that authorizes them to do so. ISO 14000 is not a "green" or "environmentally friendly" label. The certification applies to an organisation and only states that the firm has followed the steps for implementing an EMS (Wall, Weersink, and Swanton, 2001). Certain organizations believe that the ISO 14000 is irrelevant to them as it is strictly an environmental standard and many look at it as a painful process that just consumes resources. Some are yet to understand its benefits. Another myth is that it is only necessary if your products are exported to Europe (Corbett and Kirsch, 2000).

2.4 Theoretical Model

A model for the study based on the literature reviewed is presented below:

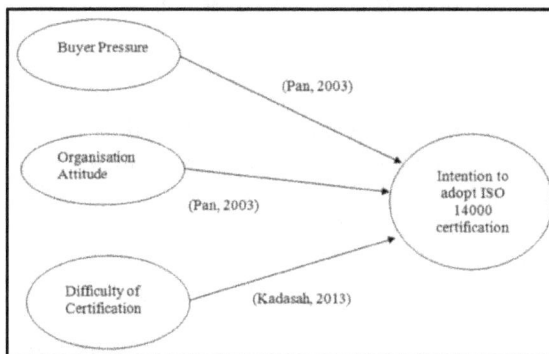

Fig. 4: Theoretical Model

As suggested by Pan (2003) and Kadasah (2013), the study will analyse the relationship between buyer presuure, oragnisation attitude and difficulty of certification on firms' intention to adopt ISO 14000 standards.

3. METHODOLOGY

3.1 Research Design

The research design was exploratory in nature. Convenience sampling was used and the sampling frame was restricted to engineering ancilliary units with less than 75 employees. Data collection involved a survey using a questionnaire. The statistical tools employed were Karlpearson's Correlation, Regression Analysis and t-tests using SPSS software.

An instrument was developed with questions to measure the three factors and the intention to adopt ISO 14000 standards. The questionnaire was given out to employees of medium and small scale enterprises in the manufacturing sector and their responses recorded. The data collection was cross-sectional rather than longitudinal as it represented a snapshot of the perceptions at one point in time. The collected data was then further analysed for other factors.

For finding out the current perceptions about ISO 14000, appointments were sought with the managers of MSMEs in the engineering ancillary sector in Coimbatore. Only those firms which had not yet implemented the ISO 14001 standard but were aware of its existence were selected. Unintentionally, a mix of small vs. big and old vs. new firms were represented.

The survey was targeted at mid to senior management employees of engineering ancilliary units with less than 75 employees in Coimbatore. Only one response per firm was accepted. The survey involved convenience (nonprobability) sampling. Managers were chosen from firms where there was a prior acquaintance with the Managers directly or with any of the employees working there. Forty four firms participated in the survey.

Karlpearson's correlation was performed to establish the magnitude and direction of relationships between buyer pressure, organisation attitude, difficulty in certification and intention to adopt ISO 14000 standards. The next step was linear regression to find out the relationship between the predictor and response variables and to come up with a possible regression equation. Finally, t-test was done to check for different significant differences among different groups of respondents and their buyer pressure levels.

3.2 Objectives

Very broadly, the study aims to find out the prevalent perceptions or opinions about ISO 14000 standards in the engineering ancillary sector in Coimbatore and also to measure the intention of MSMEs to adopt the same. The study also aims to determine the extent of influence between buyer pressure, organisation attitude, difficulty of certification and intention to adopt ISO 14000 standards.

This project has three main specific objectives:

- **Objective 1:** To study the influence of buyer pressure on intention to adopt ISO 14000 standards.

- **Objective 2:** To study the influence of organisation attitude on intention to adopt ISO 14000 standards.
- **Objective 3:** To study the influence of difficulty of certification on intention to adopt ISO 14000 standards.

4. DATA ANALYSIS

The demographic distribution of the respondents is as below:

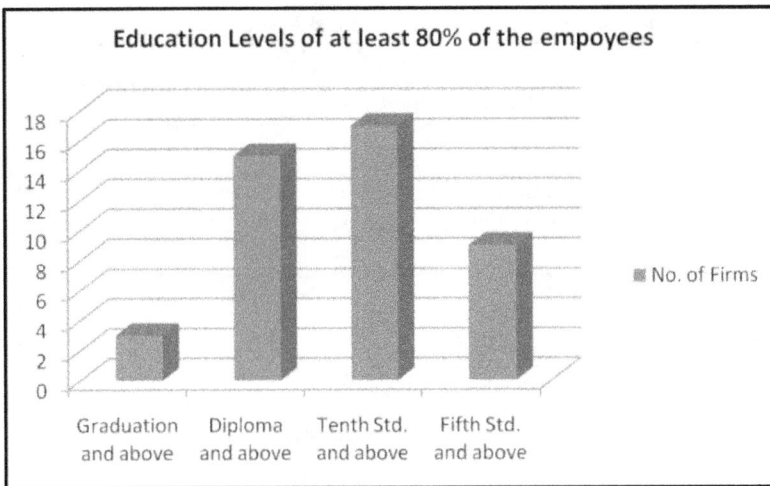

Fig. 5: Education Levels of Employees at Respondent Firms

Around forty one (93.18%) of the total number of respondent firms employed people with Diplomas or lower education levels.

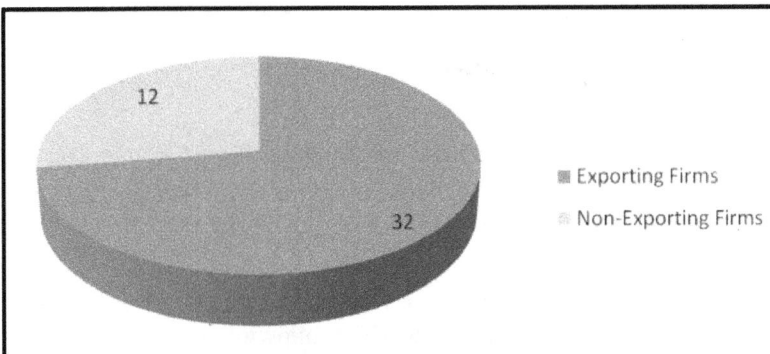

Fig. 6: Exporting *vs.* Non Exporting Firms

Thirty two (approximately 73%) of the 44 firms were exporters. The countries they exported products to included the USA, European Union and the United Kingdom, Africa and other Asian countries. The remaining firms manufactured products for local consumption.

The professional memberships of the respondent firms is given below:

Table 1: Professional Memberships of Respondent Firms

Professional Body	*No. of Firms*
No Membership	11
Coimbatore Management Association	2
Confederation of Indian Industry	5
Southern India Engineering Manufacturers' Association	4
Coimbatore District Small Industries Association	28
Indian Chamber of Commerce and Industry	2

Several firms had multiple affiliations with professional bodies. Most (63.63%) of the firms had membership with the Coimbatore District Small Scale Industries' Association (CODISSIA).

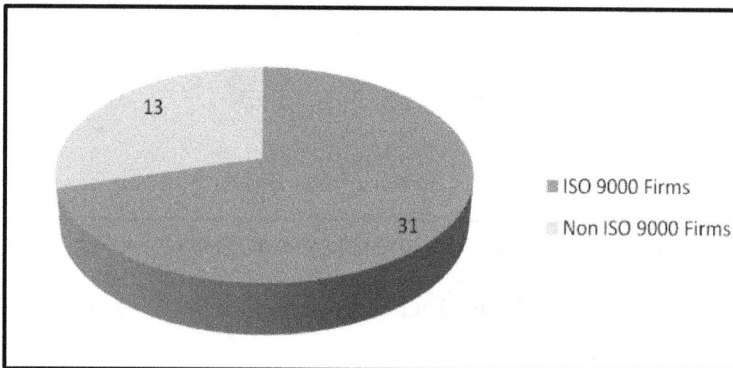

Fig. 7: Firms with ISO 9000 Certification

Thirty one (70.45%) of the 44 respondent firms were ISO 9000 certified (but not ISO 14000 certified).

Other demographic data recorded pertained to the imparting of training on environmental issues to employees. Twelve (27.27%) of the total number of respondent firms imparted environmental training at least once a year. The most favoured were informal training methods over lectures or workshops.

Based on the demographic distribution of the respondents, t-tests were conducted to test the hypotheses of equality of means for two groups. The following null hypotheses were tested:

1. H_0: There is no significant difference in buyer pressure levels between exporting and non exporting firms.
2. H_0: There is no significant difference in organization attitude levels between respondent firms that have at least 80% of their employees as diploma holders and above vs. fifth standard and above education levels.

The results obtained are shown below:

Table 2: t-Test results of Buyer Pressure

Variables	Number of Samples	Mean (Buyer Pressure)	Std. Deviation	t-value	Significance
Non –Exporting Firms	12	3.67	1.83	−6.90	0. 257
Exporting Firms	32	7.19	1.38		

Table 3: t-Test Results of Organisation Attitude

Variables	Number of Samples	Mean (Organisation Attitude)	Std. Deviation	t-value	Significance
Tenth Std and below	26	0.92	1.98	−1.62	0. 207
Diploma and above	18	1.83	1.58		

The following deductions can be drawn:

1. As the significance value (0.257) is greater than 0.05, the first null hypothesis is accepted. This implies that there is no significant difference in buyer pressure levels between exporting and non exporting firms. The reason for this could be the fact that these engineering ancillary units are low on the inbound supply chain where they enjoy very low visibility among the final customers because of which buyer pressure does not penetrate much to their level.

2. As the significance value (0.207) is greater than 0.05, the second null hypothesis is accepted. The results show that there is no significant difference in organisation attitude levels between respondent firms with at least 80% of their employees at education levels of tenth standard and below and diploma and above. This could be explained by the fact that our current system of education does not explain the adverse effects of a specific firm's manufacturing process on the environment. Our education does not provoke us to do an environmental impact analysis of any of our major decisions.

Based on their responses, individual respondent firms' buyer pressure, organisation attitude and difficulty in certification levels are depicted below:

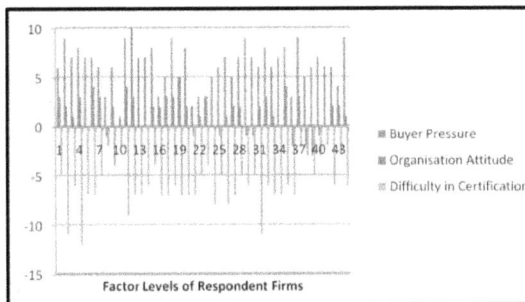

Fig. 8: Factor Levels of Respondent Firms

Influence of Factors on Intention to Adopt ISO 14000 Standards

Objective 1: Buyer Pressure and Intention to Adopt ISO 14000

Before testing for regression, a correlation test was performed to check the extent to which buyer pressure and intention to adopt ISO 14000 fluctuate with respect to each other. The test performed was a Karlpearson correlation test. The output obtained was as below:

Table 4: Correlation between Buyer Pressure and Intention to Adopt ISO 14000

		Buyer Pressure	*Intention to Adopt ISO 14000*
Buyer Pressure	Pearson Correlation	1	0.556
	Sig.		0.000
Intention to Adopt ISO 14000	Pearson Correlation	0.556	1
	Sig.	0.000	

Since the coefficient of correlation is 0.556, buyer pressure and intention to adopt ISO 14000 are positively correlated. The correlation test shows that as buyer pressure increases, intention to adopt ISO 14000 also increases.

The next test was a regression analysis to relate intention to adopt ISO 14000 (response variable) to buyer pressure (predictor variable). A linear regression analysis was performed and the following output was obtained:

Table 5: Regression Analysis for Buyer Pressure and Intention to Adopt ISO 14000

	Unstandardised Coefficients		*t*	*Sig.*
	B	*Std. Error*		
Buyer Pressure	1.383	0.756	1.829	0.074
	0.497	0.115	4.330	0.000

Based on the data above, we can create the following predictor equation:

Intention to adopt ISO 14000 = 1.383 + (0.497* Buyer Pressure)

This equation shows that even if there is no buyer pressure, there will be some intention to adopt ISO 14000. The coefficient of determination for the above test was 0.309 which implies that 69.1% of the variations are unexplained by the regression equation. The coefficient of determination signifies the strength of association or degree of closeness between the two variables. The correlation between buyer pressure and intention to adopt ISO 14000 is weak based on given data.

The weak influence of buyer pressure on intention to adopt ISO 14000 could be explained by the fact that these engineering ancillary units are low on the inbound supply chain where they enjoy very low visibility among the final customers because of which buyer pressure does not penetrate much to their level.

Objective 2: Organisation Attitude and Intention to Adopt ISO 14000

A correlation test to check the extent to which organisation attitude and intention to adopt ISO 14000 fluctuate with respect to each other was performed. The output obtained was as below:

Table 6: Correlation between Organisation Attitude and Intention to Adopt ISO 14000

		Organisation Attitude	*Intention to Adopt ISO 14000*
Organisation Attitude	Pearson Correlation	1	0.384
	Sig.		0.010
Intention to Adopt ISO 14000	Pearson Correlation	0.384	1
	Sig.	0.010	

The test showed an unexpected result. The coefficient of correlation was only 0.384 indicating a very weak correlation. A regression test here was redundant.

The lack of influence of organisation attitude on intention to adopt ISO 14000 could be explained by the fact that these engineering ancillary units are not staffed by employees who are environmentally sensitive. Neither their education nor the training offered by the organisations has prompted the employees to evaluate the impact of acquiring an ISO 14000 certification on their business.

Objective 3: Difficulty of Certification and Intention to Adopt ISO 14000

Before testing for regression, a correlation test was performed to check the extent to which difficulty of certification and intention to adopt ISO 14000 fluctuate with respect to each other. The output obtained was as below:

Table 7: Correlation between Difficulty of Certification and Intention to Adopt ISO 14000

		Difficulty of Certification	*Intention to Adopt ISO 14000*
Difficulty of Certification	Pearson Correlation	1	–0.573
	Sig.		0.000
Intention to Adopt ISO 14000	Pearson Correlation	–0.573	1
	Sig.	0.000	

Since the coefficient of correlation is –0.573, difficulty of certification and intention to adopt ISO 14000 are negatively correlated. The correlation test shows that as difficulty of certification increases, intention to adopt ISO 14000 decreases.

The next test was a regression analysis to relate intention to adopt ISO 14000 (response variable) to difficulty of certification (predictor variable). A linear regression analysis was performed and the following output was obtained:

Table 8: Regression Analysis for Difficulty of Certification and Intention to Adopt ISO 14000

	Unstandardised Coefficients		*t*	*Sig.*
	B	*Std. Error*		
Difficulty of Certification	1.579	0.684	2.308	0.026
	−0.487	0.107	−4.532	0.000

Based on the data above, we can create the following predictor equation:

Intention to adopt ISO 14000 = 1.579 − (0.487* Difficulty of Certification)

This equation shows that even if there is no difficulty of certification, there will be some intention to adopt ISO 14000. The coefficient of determination for the above test was 0.328 which implies that 67.2% of the variations are unexplained by the regression equation. The predictor equation is therefore quite weak.

The weak influence of difficulty of certification on intention to adopt ISO 14000 could be explained by the fact that most of these engineering ancillary units are already ISO 9000 certified and are aware of at least some of the processes and cost and time implications of the ISO 14000 certification process on their business. Had the process been mandatory, they would have undergone the process.

5. FINDINGS

The following findings were obtained:

- There is no significant difference in buyer pressure levels between exporting and non exporting firms.
- There is no significant difference in organization attitude levels between respondent firms that have at least 80% of their employees as diploma holders and above vs. fifth standard and above education levels.
- The correlation between buyer pressure and intention to adopt ISO 14000 levels is positive but weak.
- The correlation between organisation attitude and intention to adopt ISO 14000 levels is extremely weak.
- The correlation between difficulty in certification and intention to adopt ISO 14000 levels is negative but weak.

The findings imply the following:

1. The engineering ancillary units are low on the inbound supply chain where they enjoy very low visibility among the final customers because of which buyer pressure does not penetrate much to their level.
2. Our current system of education does not explain the adverse effects of a specific firm's manufacturing process on the environment. Our education does not provoke us to do an environmental impact analysis of any of our major decisions.

3. The engineering ancillary units are not staffed by employees who are environmentally sensitive. Neither their education nor the training offered by the organisations has prompted the employees to evaluate the impact of acquiring an ISO 14000 certification on their business.

4. Most of these engineering ancillary units are already ISO 9000 certified and are aware of at least some of the processes and cost and time implications of the ISO 14000 certification process on their business. Had the process been mandatory, they would have undergone the process.

6. SCOPE

Although there are several drivers for adoption of ISO 14000 standards that can be measured, this study concentrates on the measurement of buyer pressure, oraganisation attitude and difficulty of certification and the extent of influence of these three factors on the intention to adopt ISO 14000 standards.

The respondents surveyed belonged to the micro and small scale enterprises in the manufacturing industry but were not restricted to any functional capacity or age group or management level. The firms chosen did not have ISO 14000 certification but were aware of the existence of such standards.

7. LIMITATIONS

The study has several limitations:

1. The respondents were from a variety of different levels in organizations. The data gathered is dependent on the respondent's level of knowledge in ISO 14000 standards and its implementation.

2. Only three relationships, i.e. impact of buyer pressure, organization attitude and difficulty in certification on intention to adopt ISO 14000 certification were investigated.

3. This study does not deal with the differences in the scope and reach of the widely popular ISO 9000 standards and the less popular ISO 14000 standards.

8. CONCLUSION

ISO 14000 family of environmental standards are just a means to help us implement an environmental management system that is suitable and customized to our industry. There are several drivers which promote and act as barriers to the successful implementation of the ISO 14001 standard in any organisation.

The analyses of the data by the respondents in the engineering ancillary sector in Coimbatore suggests that buyer pressure is positively but weakly correlated to the intention to adopt ISO 14000 standards. There is no influence of organisation attitude on intention to adopt ISO 14000 standards. Difficulty in certification is weakly negatively correlated to the intention to adopt ISO 14000 standards. These findings are in line with the fact that these organsations are not ISO 14000 certified and that they do not have any plans to undergo the certification process in the near future.

The reasons for these findings may be impacted by the number of respondent firms and the knowledge level of responding managers. A further study with an increased number of respondents more focused on producing one type/variety of product may be necessary.

REFERENCES

Arvanitoyannis, I.S. and Boudouropoulos, I.D. (1999). Current state and advances in the implementation of ISO 14000 by the food industry—Comparison of ISO 14000 to ISO 9000 to other environmental programs. *Trends in Food Science and Technology*, 9, 395–408.

Ball, J. (2002). Can ISO 14000 and eco-labelling turn the construction industry green? *Building and Environment*, 37(4), 421–428. doi:10.1016/S0360-1323(01)00031-2.

Cascio, J. and Baughn, K.T. (2000). Health, Safety and ISO 14001. *Manufacturing Engineering*, 124 (December), 126–135.

Cassells, S., Findlater, A. and Lewis, K. (2009). 2008 International Council for Small Business World Conference, 1–17.

Cassells, S., Lewis, K.V. and Findlater, A. (2012). An exploration of ISO 14001 uptake by New Zealand firms. *International Journal of Law and Management*, 54(5), 345–363. doi:10.1108/17542431211264232.

Corbett, C.J. and Kirsch, D.A. (2000). ISO 14000: An agnostic's report from the front line. *International Organization for Standardization - ISO. In ISO 9000 + ISO 14000 News, March–April*, 4–17.

Curkovic, S., Sroufe, R., Melnyk, S. and Montabon, F.L. (2001). Identifying the Factors which Affect the Decision to Attain ISO 14000. In *Supply Chain and Information Management Conference Papers, Posters and Proceedings*. Retrieved from http://lib.dr.iastate.edu/scm_conf/4.

Elefsiniotis, P. and Wareham, D.G. (2005). ISO 14000 Environmental Management Standards: Their Relation to Sustainability. *Journal of Professional Issues in Engineering Education and Practice* (July), 208–212.

Hariz, S. and Bahmed, L. (2013). Assessment of environmental management system performance in the Algerian companies certified ISO 14001. *Management of Environmental Quality: An International Journal*, 24(2), 228–243. doi:10.1108/14777831311303100.

Haufler, V. Negotiating International Standards for Environmental Management Systems: The ISO 14000 STANDARDS. Retrieved from www.globalpublicpolicy.net.

ISO Guidance for Small and Medium-Sized Businesses. (2006). *Business and Environment*, 17(11), 12.

Kadasah, N.A. (2013). Attitudes of Managers towards the Potential Effects of ISO 14001 in Saudi Arabia: Factor Analysis. *International Business Research*, 6(7), 91–102. doi:10.5539/ibr.v6n7p91.

Marett, G.E. (2000). *Adoption of the ISO 14001 Environmental Management System Standard: Implications for Agri-Business*. Dalhousie University, Canada.

Marimon, F., Heras, I. and Casadesús, M. (2009). ISO 9000 and ISO 14000 standards: A projection model for the decline phase. *Total Quality Management and Business Excellence*, 20(1), 1–21. doi:10.1080/14783360802614257.

Murray, P.C. (1997). International Environmental Management Standard, ISO 14000: A Non-Tariff Barrier or a Step to an Emerging Global Environmental Policy. *Journal of International Law*, 18(2), 577. Retrieved from http://scholarship.law.upenn.edu/jil/vol18/iss2/7.

Pan, J. (2003). A comparative study on motivation for and experience with ISO 9000 and ISO 14000 certification among Far Eastern countries. *Industrial Management and Data Systems*, 103(8), 564–578. doi:10.1108/02635570310497611.

Pawliczek, A. and Piszczur, R. (2013). Consequence of QMS ISO 9000 and EMSISO 14000 Implementation on CZ/ SK Enterprise Performance with Respect to Sustainability. *Journal of Eastern Europe Research in Business and Economics*, 2013, 1–18. doi:10.5171/2013.420617.

Pouliot, C. (1996). ISO 14000: Beyond Compliance to Competitiveness. *Manufacturing Engineering*, 116 (December), 51–56.

Ratiu, P. and Maria, M. (2014). Dynamics of Certified Environmental Management Systems: ISO 14001 AND EMAS IN ROMANIA. *Annales Universitatis Apulensis Series Oeconomica*, 16(1), 198–211.

Searcy, C., Morali, O., Karapetrovic, S., Wichuk, K., McCartney, D., McLeod, S. and Fraser, D. (2012). Challenges in implementing a functional ISO 14001 environmental management system. *International Journal of Quality and Reliability Management*, 29(7), 779–796. doi:10.1108/02656711211258526.

Strachan, P.A., Sinclair, I.M. and Lal, D. (2003). Managing ISO 14001 implementation in the United Kingdom Continental Shelf (UKCS). *Corporate Social Responsibility and Environmental Management*, 10(1), 50–63. doi:10.1002/csr.29.

Susan, L.K. (2007). ISO 14001 Hits 10-Year Mark. *Quality Progress*, 40(8), 67.

Vastag, G. (2003). Revisiting ISO 14000 Diffusion: A New " Look" at the Drivers of Certification. *Production and Operations Management*, (October).

Wahid, N.A. and Y.N.G. (2010). The Effect of ISO 14001 Environmental Management System Implementation on SMEs Performance: An Empirical Study in Malaysia. *Journal of Sustainable Development*, 3(2), 215–221.

Wall, E., Weersink, A. and Swanton, C. (2001). Agriculture and ISO 14000. *Food Policy*, 26(1), 35–48. doi:10.1016/S0306-9192(00)00025-7.

About ISO - ISO. (n.d.). Retrieved from http://www.iso.org/iso/home/about.htm Last accessed 6 December 2014.

ISO 14000 - Environmental management - ISO. (n.d.). Retrieved from http://www.iso.org/iso/home/standards/management-standards/iso14000.htm Last accessed 6 December 2014.

Impact of Social Media on Management Students for Social Service and Sustainability

Vinith Kumar Nair[1], Arya Pinto[2], Lakshmi Chandran[3] and Sneha Satheesh[4]
[1]TKM Institute of Management, Kerala
[2–4]TKM Institute of Management, Kollam–691505, Kerala
E-mail: [1]vinith79@gmail.com; [2]aryapinto1993@gmail.com;
[3]lakshmichsu@gmail.com; [4]snehasatheesh70@gmail.com

ABSTRACT: *The purpose of the study was to examine the adoption of social service activities for sustainability by the management students in Kerala using social media. The study also defines the nature of adolescent social behavior as revealed through activities on social media that contributes to sustainability. The study was conducted among 350 randomly selected students from top 5 management institutes in Kerala. A validated Questionnaire was used for the study. Using SPSS the collected data was analyzed. This study found that influence of social media can be associated with social service and sustainability among management students. The findings of this study imply that social media can be used to sensitize about importance of social work and that will help in bringing sustainable thinking in decision making. The focus on social service activities through social media will release enormous opportunities for social workers to bring in sustainable approach. The management graduates through the use of social media can clearly understand and analyze the sustainability frameworks and practices which every business faces in becoming sustainable. Sustainability is critical to future success of corporate, thus training focusing on these aspects can be provided to the next generation of leaders to deal with sustainability Social media can become a new interface tool for the same. Social media can be used to crowd source ideas to find innovative solutions to social service activities and sustainability dilemmas.*

Keywords: Social Media, Social Services, Sustainability, Sustainable Development, Corporate Sustainability, Corporate Social Sustainability.

1. INTRODUCTION

Social media is an on-line technology which enables people to communicate and instantly share information and resources with local, national and international audiences. It includes, Facebook, Twitter, LinkedIn, Google+, Skype, YouTube, on-line blogs, etc. Andres Kaplan (2010) described in his study that "social media is a set of internet based application that constructs on the ideological and technological foundation of web and that permit the design and exchange of user generated content." As social media use continues to evolve and expand, social workers must examine the use of this technology within the domain of professional practice and ethical decision-making. It is also cheapest and which gives fastest access to the world. The use of social media is increasing day by day all over the world. Research reports have suggested that its influences are high on youth.

"Youth is the time of life when someone is young. Youth is the time when a young person has not yet become an adult. Youth is very important for future of any nation and country's progress and development. Now a day Social media is essential for youth in the field of education to learn new trends in education, to improve writing and communicating skills, cultural promoting, religious and political information gathering and sharing links, better living style, growth and development of society (Merriam Encyclopedia, 2001).

Society is currently immersed in a deep economic, social, environmental and political crisis. In the face of this crisis an increasingly strong model for sustainable societal development grounded in human rights and allowing for a fairer world is the need of the hour. It is based on a commitment to solidarity with both present and future generations, ensuring itsdurability over time and it does not merely focus on economic growth as the key to development (Major Zaragoza, 2009).

"IFSW calls on social workers to recognize the importance of natural and built environment to the social environment, to develop environmental responsibility and care for the environment in socialwork practice and management today and for future generations, to work with other professionals to increase our knowledge and with community groups to develop advocacy skills and strategies towork towards a healthier environment and to ensure that environmental issues gain increased presence in social work education" (International Policy Statement on Globalisation and Environment).

Countries like India consider social service to be the best form of God's prayer and all the religions believes in the same directions. Social service is the need of present time, as we don't live for ourselves but for the society we are living in. Today's Indian youth is sensible and capable enough to sensitize the issues and fight for the betterment of society in all manners. With the communication revolution everyone is connected to thousands and lakhs of acquaintances. With just one click we can share whatever we want to communicate and this is the power of social media. Social Media can be used as a tool to promote social service and sustainability.

Programmes like Satyamev Jayate, a TV show discusses and provides possible solutions to address social issues in India. The show focus on sensitive social issues prevalent in India such as female foeticide, child sexual abuse, rape, honour killings, domestic violence, untouchability, alcoholism, and the criminalization of politics. This programme aimed to encourage citizens to take action by providing information about their country.Another such initiative is JWALA, a well known Charitable Trust situated in a rented villa at the capital city of Kerala, Trivandrum with an aim to "bringing together the poor and needy those living on the streets" has formed in its shape in the year 2011. The UW Recycling Facebook page has become a main pipeline to speak with students and staff and to raise awareness about campus-wide efforts towards sustainability and waste reduction and there are many such efforts taken through social media to spread awareness among youths.

Sustainability is one of the most revolutionarily social perspectives that emerged in the 20th century. When human beings began to realize that their actions were eating away the fragile environment on which their own lives depend, this new perspective got emerged. The newly emerged social perspective was widely accepted in many respects since it is allied to almost all aspects of modern day living. Thus it offers a new lens, a new way of understanding reality. Hence, it may be called a paradigm. In addition, it calls for determined and consistent changes in our relationship with the environment. The popular understanding of sustainability as an environmental phenomenon was a result of the immediate connection with the physical environment during the early days of the emergence of the sustainability paradigm.

Almost always, the term sustainability invokes in one's mind pictures of ominous environmental disasters. As a result, scholars, practitioners, and advocates from the environmental and physical sciences have had an enormous influence on the field of sustainability. However, lately a new awareness is emerging that "environment is related to all other facets such as the social and economic aspects of life. This view has led to the legitimate emergence of social scientists as one group among several stakeholders in achieving sustainability" (Paehlke, 2001).

The recognition of the equality among social, economic, and physical environments has several implications (Kondrat, 2002). The Brundtland Report, titled "Our Common Future, places ecological sustainability on an equal footing with social and economic sustainability. Massive consumption of world renewable and non-renewable resources in countries of the North was noted as a contributor to the environmental degradation. Recognition that both population growth and consumption can have deleterious effects on the environment began to influence the sustainability debate" (United Nations, 1987).

"The Rio summit called for a global partnership for sustainable development especially through poverty reduction. Poverty is the biggest polluter. It pollutes systems at all levels, social, economic, and environmental." (Glasmeier and Farrigan, 2003). "It became clear that environmental degradation cannot be stopped unless poverty and social inequalities created by market and nonmarket forces are addressed through poverty reduction measures It also became clear that economic development in most developing countries was also followed by increase in social inequality" (Carrilio, 2007).

The development of social work as a meaningful and viable option requires changes and transformation in paradigm, social policy and social services. This will facilitate a sustainable future for all mankind.

2. LITERATURE REVIEW

2.1 Social Media

"Social media technologies that facilitate social interaction, make possible collaboration, and enable deliberation by stake holders across boundaries, time and space. These technologies include blogs, wikis, media (audio, photo, video, text) sharing tools, networking platforms (including facebook), and virtual words" (Kaplan and Haenlein, 2010). Social Media has the ability to communicate, share information and resources with a wider audience at low cost. Social media sites are being used by social workers in their professional and personal lives.

Word of mouth is the key aspect of social media; it has innumerable communication and conversions between the involved parties. People talk about various topics and issues, one of the issues is social service (Kotler, 2009).

2.2 Social Service

"Social work is a practice-based profession and an academic discipline that promotes social change and development, social cohesion, and the empowerment and liberation of people. Principles of social justice, human rights, collective responsibility and respect for diversities are central to social work. Underpinned by theories of social work, social sciences, humanities and indigenous knowledge, social work engages people and structures to address life challenges and enhance wellbeing" (IFSW General Meeting and the IASSW General Assembly, July 2014). Social service is an activity designed to promote social well-being. It include facilities such as education, food subsidies, health care and subsidized housing provided by a government to improve the life and living conditions of the children, disabled, the elderly and the poor in the national community.

While use of internet media is considered as a challenge in tobacco control policy, a study done by Chawada (2013) favours use of social media as a communication tool to spread messages among the internet users (Mackenzie, 2012).

2.3 Sustainability

Sustainability is "the result of the growing awareness of the global links between mounting environmental problems, socio-economic issues to do with poverty and inequality and concerns about a healthy future for humanity" (Hopwood *et al.,* 2005). It emphasise on meeting the needs of the present without compromising the ability of future generations to meet their own needs. Sustainability is therefore the protection and well-being of the earth's natural cycles, millions of species of plants and animals, including humankind.

2.4 Sustainable Development

Sustainable development is 'maintaining a positive process of social change' (Baker, 2006) with a global perspective that concerns the needs of people and, therefore, takes account of the dimensions of space–where they are located–and the historical junctures in their location's development. "Sustainable development is development that meets the needs of the present without compromising the ability of future generations to meet their own needs. It contains within it two key concepts: the concept of 'needs', in particular the essential needs of the world's poor, to which priority should be given; and the idea of limitations imposed by the state of technology and social organization on the environment's ability to meet present and future needs (World Commission on Environment and Development, WCED, 1987).

5.5 Corporate Sustainability

Corporate sustainability is "an approach to enhance competitive position by taking opportunities and managing sustainability risks drawn from global trends to ensure that the needs of direct and indirect stakeholders will be met today and in future" (Will, 2008). In this resource constrained world, the principles of sustainable development align a company's decision-making about the allocation of capital, product development, brand and sourcing.

2.6 Social Sustainability

"Social Sustainability for a city is defined as development which is compatible with the harmonious evolution of civil society, fostering an environment conductive to the compatible contribution of culturally and socially diverse groups while at the same time encouraging social integration with improvements in the quality of life of all segments of the population." (UNESCO).

"Social sustainability refers to the sustenance of basic human needs such as nutrition and shelter (Streeten, Burki, Haq, Hicks, and Steward, 1981); human freedoms, including political rights, economic facilities, social opportunities, transparency guarantees, and protective security (Sen, 1999); and human development, which expands social, economic, cultural, and political choices and leads to equity sustainability, productivity, and empowerment". Thus, social sustainability enables human beings to satisfy their basic needs and achieve a reasonable standard of living.

"Social sustainability is the progress toward enabling all human beings to satisfy their essential needs; to achieve a reasonable level of comfort; to live lives of meaning and interest; and to share fairly in opportunities for health and education". (Assefa and Frostell, Chiu, Jennings and Zandbergen, 2007, 2003, 1995)

2.7 Corporate Social Sustainability

"Corporate social sustainability is encompassing strategies and practices that aim to meet the needs of the stakeholders today, while seeking to protect, support, and enhance the human and natural resources that will be needed in the future." (Thomas Dyllick and Kai). Thus the business is obligated to behave ethically and contribute to economic development by remodelling the standard of living of the work force and their families as well as a local community and society at large.

2.8 Social Media and Social Service

"Social media apps offer social workers powerful aids to their practices, but new ethical dilemmas, as well. Practitioners are figuring out how to utilize programs such as Facebook, LinkedIn, Twitter, and Web-based blogs in ways that are consistent with the long-standing rules of their areas of practice—and they don't all agree on how to go about it" (Kathryn Chernack, 2013).

According to Lenhart *et al.* (2010), "about 57% of social network users are 18–29 years old and have a personal profile on multiple social media websites. Students spent an average of 47 minutes a day on Facebook". "More than 50% of college students go on a social networking site several times a day" (Sheldon, 2008). Quan-Haase and Young (2010) found that "82% of college students reported logging into Facebook several times a day. Younger students tended to use Facebook more frequently than older students to keep in touch with friends from high school or from their hometown" (Pempek *et al.,* 2009).

Hypothesis 1: Impact of Social Media is having significant relationship with Social Service.

Hypothesis 2: Age of Management Students is having significant relationship with use of Social Media.

Hypothesis 3: Gender of Management Students is having significant relationship with use of Social Media.

2.9 Social Service and Sustainability

Peeters (2012) explains that there is a meaningful normative similarity between the principles of sustainable development in its social dimension and social work, in terms of attention to welfare, intergenerational fairness, the defence of human rights and social justice, and the promotion of gender equality, diversity and active citizenship. This similarity is grounded in the very definition of social work according to the IFSW and in its ethical principles. So, if a social worker accomplishes a project forthe betterment of homeless persons, they must take into account that such action cannot damage the environment. They cannot create housing for all the homeless persons through deforestation.

If students have a poor understanding on a particular social issue, they tend to rely on the information provided by the social media (Thogersen, 2006). The UK Sustainable Development Strategy identifies "a need to make 'sustainability literacy' a core competency for professional graduates'. All graduates will share responsibility as stewards, not only of the environment, but also of social justice–as employees, citizens and, in many cases, parents and mentors of the next generation". The management graduates are the future corporate, so, they should be aware about value of social service and they should be encouraged to serve the people of their country as well as to protect the environment for their future generations.

The sustainability paradigm even though it came into light due to the increasing trend of human made crisis and some of the paradigm existed within the arena of social work. As a helping hand social workers mainly the management students where part of the interventions. The systematic approach to the concern of sustainability has led to advances in the field of social work. Thus, it's believed that if sustainability is a new paradigm, social work or social service as a discipline provides the platform for sustainable thinking and actions.

Hypothesis 4: Social Service is having significant relationship with Sustainability.

Hypothesis 5: Age of Management Students is having significant relationship with Social Service.

Hypothesis 6: Gender of Management Students is having significant relationship with Social Service.

2.10 Social Media and Sustainability

Dyllick and Hockerts, 2002 describe social sustainability on a corporate level and define that "socially sustainable companies add value to the communities within which they operate by increasing the human capital of individual partners as well as furthering the societal capital of these communities. They manage social capital in such a way that stakeholders can understand its motivations and can broadly agree with the company's value system."

Social media has undeniably changed the corporate sustainability landscape as companies harness this technology for their sustainability programs. According to the SMI-Wizness Social Media Sustainability Index, the number of companies in its index that use social media to communicate sustainability has more than doubled in the last year (Kimberly Wilson, 2012) "Social media can be both a carrot and a stick. The new media serves the growth of sustainability as a powerful communications tool and as a persuasive means of pushing for pro-social change. There are many people within a company, and groups interested in the company, who would welcome the sustainability engagement via social media. When social media and sustainability work together, sustainability becomes interesting and accessible (Richard Mathews, 2013). Therefore, the following hypothesis is constructed for analytical examination.

Hypothesis 7: Impact of Social Media is having significant relationship with Sustainability.

Hypothesis 8: Age of Management Students is having significant relationship with focus towards Sustainability.

Hypothesis 9: Gender of Management Students is having significant relationship with Sustainability.

3. METHODOLOGY

The study, descriptive in nature, was conducted among the management graduates of top 5 management institutes in Kerala. There were 350 respondents who were randomly selected for the study. Using a multistage sampling technique the respondents were identified from different management institutes. The respondents in this study consist of both first year and second year MBA graduates. Different factors for the study were identified on the basis of literature review. Two rounds of focus group discussion each consisting of 10 management students giving equal representation to both male and female participants from both first year and second year were conducted to find out the validity of the factors identified for the study. Focus group discussion is a form of qualitative research in which a group of people gather together to discuss a specific topic of interest. The members of the group share their perception and opinion related to topic being discussed. Once the factors were identified, it was shown to a panel of experts which included academicians, researchers and social workers and after a final review a validated Questionnaire was used for the study. The data was collected from only those respondents who use social media. The questionnaire was enclosed with a cover letter explaining the purpose of the study, brief objectives of the study and ensuring the respondents of the confidentiality of their responses.

4. METHOD OF ANALYSIS

Data collected through questionnaire was classified, coded, tabulated and analysed with the help of Statistical Package for Social Sciences (SPSS). Age and gender of management students were taken to analyse the influence of social media for social service and sustainability among them. Suitable hypothesis were formulated Parametric test was adopted to establish a relationship between the variables. The variables are measured using five-point Likert scale varying from (1) Strongly disagree (2) Disagree (3) Neither disagree nor agree (4) Agree (5) Strongly agree. Chronbach's Alpha was used to test the internal consistency and reliability of the scale. Karl Pearson's correlation test and independent t test was used to test the hypothesis.

5. ANALYSIS AND DISCUSSION

Table 1

Reliability Statistics

Cronbach's Alpha	Cronbach's Alpha Based on Standardized Items	No. of Items
.896	.899	20

The results in Table 1 presents that the Cronbach's Alpha values of the construct ranged from .896 which signifies that the construct has acceptable internal consistency. The result of Construct Reliability (CR) value is .896 which exceeds the expected threshold .70. The internal consistency of values indicates that convergent validity is supported.

Table 2

Correlations

		Social Media	Social Service
Social Media	Pearson Correlation	1	.579**
	Sig. (2-tailed)		.000
	N	350	350
Social Service	Pearson Correlation	.579**	1
	Sig. (2-tailed)	.000	
	N	350	350

**. Correlation is significant at the 0.01 level (2-tailed).

Table 2 represents the correlation between social media and social service.

- *H0a:* There is no significant relationship between social media and social service.
- *H1a:* There is significant relationship between social media and social service.

There is significant positive correlation between social media and social service with a significant value of 0.579. Hence the alternate hypothesis that there is significant relationship between social media and social service is accepted.

Table 3

Correlations

		Social media	Sustainability
Social Media	Pearson Correlation	1	.650**
	Sig. (2-tailed)		.000
	N	350	350
Sustainability	Pearson Correlation	.650**	1
	Sig. (2-tailed)	.000	
	N	350	350

**. Correlation is significant at the 0.01 level (2-tailed).

Table 3 represents the correlation between social media and sustainability.

- *H0b:* There is no significant relationship between social media and sustainability.
- *H1b:* There is significant relationship between social media and sustainability.

There is significant positive correlation between social media and sustainability with a significant value of 0.650. Hence the null hypothesis is rejected.

Table 4

Correlations

		Social Service	Sustainability
Social Service	Pearson Correlation	1	.447**
	Sig. (2-tailed)		.000
	N	350	350
Sustainability	Pearson Correlation	.447**	1
	Sig. (2-tailed)	.000	
	N	350	350

**. Correlation is significant at the 0.01 level (2-tailed).

Table 4 represents the correlation between social service and sustainability.

- *H0c:* There is no significant relationship between social service and sustainability.
- *H1c:* There is significant relationship between social service and sustainability.

There is significant positive correlation between social service and sustainability with a significant value of 0.447. Hence the alternate hypothesis that there is significant relationship between social service and sustainability is accepted.

Table 5

Correlations

		Age	Social Media
Age	Pearson Correlation	1	.021**
	Sig. (2-tailed)		.690
	N	350	350
Social Media	Pearson Correlation	.021**	1
	Sig. (2-tailed)	.690	
	N	350	350

**. Correlation is significant at the 0.01 level (2-tailed).

Table 5 represents the correlation between social media and social service.

- *H0d:* There is no significant relationship between age and social media.
- *H1d:* There is significant relationship between age and social media.

Since, the significant level (r = 0.021) is lesser than the acceptable level 0.05, this analysis accept the null hypothesis. Hence there is no significant relationship between age and social media.

Table 6

Correlations

		Age	Social Service
Age	Pearson Correlation	1	$-.176^{**}$
	Sig. (2-tailed)		.001
	N	350	350
Social Service	Pearson Correlation	$-.176^{**}$	1
	Sig. (2-tailed)	.001	
	N	350	350

**. Correlation is significant at the 0.01 level (2-tailed).

Table 6 represents the correlation between age and social service.

- *H0e:* There is no significant relationship between age and social service.
- *H1e:* There is significant relationship between age and social service.

Since, the significant level (r = –0.176) is lesser than the acceptable level 0.05, this analysis rejects the null hypothesis. Hence there is small relationship between age and social service.

Table 7

Correlations

		Age	Sustainability
Age	Pearson Correlation	1	-.103**
	Sig. (2-tailed)		.055
	N	350	350
Sustainability	Pearson Correlation	−.103**	1
	Sig. (2-tailed)	.055	
	N	350	350

**. Correlation is significant at the 0.01 level (2-tailed).

Table 7 represents the correlation between age and sustainability.

- *H0f:* There is no significant relationship between age and sustainability.
- *H1f:* There is significant relationship between age and sustainability.

Since, the significant level (r = −0.103) is greater than the acceptable level 0.05, this analysis accept the null hypothesis. Hence there is a significant relationship between age and sustainability.

Table 8

Independent Samples Test

		Levene's Test for Equality of Variances		t-Test for Equality of Means						
		F	Sig.	T	Df	Sig. (2-tailed)	Mean Difference	Std. Error Difference	95% Confidence Interval of the Difference	
									Lower	Upper
Social Media	Equal Variances Assumed	.493	.483	−1.095	348	.274	−.548	.501	−1.533	.436
	Equal Variances not Assumed			−1.095	347.995	.274	−.548	.501	−1.533	.436

The above table indicates the mean and standard error between gender of management students and social media.

- *H0g:* There is no significant difference between gender of management students and social media
- *H1g:* There is significant difference between gender of management students and social media.

The mean difference is −.548 for male and female and standard error difference is 0.501 while t value is −1.095. The significant level (p = 0.274) is greater than the acceptable level of 0.05, the analysis failed to reject the null hypothesis. Hence, there is no significant difference between gender of management students and social media.

Table 9

Independent Samples Test

		Levene's Test for Equality of Variances		t-test for Equality of Means						
		F	*Sig.*	*t*	*Df*	*Sig. (2-tailed)*	*Mean Difference*	*Std. Error Difference*	*95% Confidence Interval of the Difference*	
									Lower	*Upper*
Social Service	Equal Variances Assumed	.705	.402	−.475	348	.635	−.048	.100	−.244	.149
	Equal Variances not Assumed			−.475	347.958	.635	−.048	.100	−.244	.149

The above table indicates the mean and standard error between gender of management students and social service.

- *H0h:* There is no significant difference between gender of management students and social service
- *H1h:* There is significant difference between gender of management students and social service.

The mean difference is −0.048 for male and female and standard error difference is 0.100 while t value is −0.475. The significant level (p = 0.635) is greater than the acceptable level of 0.05, the analysis failed to reject the null hypothesis. Hence, there is no significant difference between gender of management students and social service.

Table 10

Independent Samples Test

		Levene's Test for Equality of Variances		t-test for Equality of Means						
		F	*Sig.*	*T*	*Df*	*Sig. (2-tailed)*	*Mean Difference*	*Std. Error Difference*	*95% Confidence Interval of the Difference*	
									Lower	*Upper*
Sustainability	Equal Variances Assumed	1.199	.274	−.920	348	.358	−.273	.297	−.856	.310
	Equal Variances not Assumed			−.920	347.188	.358	−.273	.297	−.856	.311

The above table indicates the mean and standard deviation between gender of management students and social sustainability.

- *H0i:* There is no significant difference between gender of management students and sustainability

- *H1i:* There is significant difference between gender of management students and sustainability.

The mean difference is –0.273 for male and female and standard error difference is 0.297 while t value is –0.920. The significant level (p = 0.358) is greater than the acceptable level of 0.05, the analysis failed to reject the null hypothesis. Hence, there is no significant difference between gender of management students and sustainability.

6. DISCUSSION

This study sought to establish a baseline of descriptive information on the usage of social media among management students and their attitude on social service and sustainability. The findings of the present study have made it abundantly clear that there is significant positive correlation between social media usage, social service and sustainability.

The result of this study shows that there is no significant relationship between age of management students and social media and very small relationship between age and social service. There is significant relationship between age and sustainability. The study also highlights the fact that there is no significant difference between gender of management students and social media, social service and sustainability. The implication of the findings of this study explains that social media can be used to sensitize about importance of social work and that will help in bringing sustainable thinking in decision making.

Social service has a great role in developing sustainable future. Sustainability is critical to future corporate success, thus training focusing on these aspects can be provided to the next generation of leaders to deal with sustainability and social media can become a new interface tool for the same. The social media has an immense role as an essential digital tool for the leaders in social services to reach wider audience particularly the students by conveying them about current environmental situations. This encourages them to work for the betterment of the society. While considering sustainability, the efforts depends utmost with the society and the youth which in this case is management student's. This encourages the students for social service activities that ultimately leads to sustainability. Different social media platforms will help to attain sustainability with the platforms such as Facebook allowing the students to exchange messages on different social service activities and which ultimately inspires them to work towards the change in the current environment.

The contents of sites like Facebook create educational experiences for the students and acts as catalyst for change. Thus, the research highlights the point that social media effectively acts as a suitable medium to spread awareness on the subject of social services for sustainability among the management students. It meets the need of the audience mainly the management students and acts as a reliable source for them to get updated of their physical surrounding and respond to the cause. The success of the Social Media depends to a great extent on the contents presented in the media. So social service activities can get a boost if social media is extensively used to sensitise management students and which in turn will lead to sustainability. Thus, proper social media content creation is important for drawing the students so as to encourage them and thereby focusing to a great extent to contribute to social services for sustainability in order to maintain credibility.

7. CONCLUSION

The study highlights the fact that there is ample scope in exploring social media in communicating about different aspects of social work and sustainability among management students. Organisations engaging in social work should leverage the use of social media to sensitise the management students who irrespective of their age and gender extensively use it about social work and sustainability. Since, limited studies are conducted in this area of influence of social media on social services for sustainability and particularly in the Indian context, this study will open the gates for future studies.

REFERENCES

Andres Kaplan (2010). Human Development and Economic Sustainability. World Development, 28(12), 2029–2049.

Assefa and Frostell, Chiu, Jennings and Zandbergen (2007, 2003, 1995). Information for clinical social work practice: A potential solution. Clinical Social Work Journal, 40, 166–174.

Assefa and Frostell, Chiu, Jennings and Zandbergen (2007, 2003, 1995). Planning for sustainable development: A paradigm shift towards a process based approach Sustainable Development, 15(2), 83–89.

Baker (2006). Complaining to the masses: The role of protest framing in customer-created complaint web sites. *Journal of Consumer Research*, 33(2), 220–230.

Carrilio, T.E. (2007). Utilizing a social work perspective to enhance sustainable development efforts in Loreto, Mexico. International Social Work, 50(4), 528–538.

Carrilio, T.E. (2007). Utilizing a social work perspective to enhance sustainable development efforts in Loreto, Mexico. *International Social Work*, 50(4), 528–538.

Chawada (2013). A three-dimensional Conceptual Model of Corporate Social Performance. In: *Academy of Management Review,* Vol. 4, 4, pp. 497–505.

Dyllick and Hockets (2002). Proposal for sustainable development goals. Retrieved from website https://sustainabledevelopment.un.org/focussdgs.htm

Glasmeier, A.K. and Farrigan, T.L. (2003). Poverty, sustainability, and the culture of despair: Can sustainable development strategies support poverty alleviation in America's most environmentally challenged communities? Annals of the American Academy of Political and Social Science, 590, 131–149.

Haq. (2000). Social dimensions of the environmental crisis: Challenges for social work. Knowledge systems for sustainable development. *Proceedings of the National Academy of Sciences*, 100(14), 8086–8091.

Hopwood *et al. (*2005). Promoting healthy lifestyle for sustainable development. In: International Journal of Sustainable Development, Vol. 7, 1, pp. 59–75.

Kaplan, A.M. and Haenlein, M. (2010). The fairyland of Second Life: About virtual social worlds and how to use them. Business Horizons, 52(6), 563–572.

Kathryn Chernack (2013). Conclusion: implications for management practice, education, and research. In N. Roome (Ed.), Sustainability Strategies for Industry. The Future of Corporate Practice (pp. 259–276). Washington DC: Island Press, p. 275.

Kathryn Chernack (2013). Queen, International of Journal of Sustainable Development, 6(4), pp. 395–416.

Kimberly Wilson (2012). Empowerment as guidance for professional social work: An act of balancing on a slack rope. *European Journal of Social Work*, 6(3), 229–240.

Kondrat (2002). Ethical consequences of using social network sites for students in professional social work programs. Journal of Social Work Values and Ethics, 9(1), 5–8.(2009d). Learning Pathway "on ACMA's Cyber Smart Learning Pathways".

(2009b), Click and Connect: Young Australians" Use of Online Social Media – Pt2 Australian Communications and Media Authority, Canberra.

(2009c), Developments in internet filtering technologies and other measures for promoting online safety the Second Annual Report to the Minister for Broadband, Communication and the Digital Economy.

Kotler (2009). Social Media as a tool of marketing and creating. Retrieved from http://social-media-monitoring-review.toptenreviews.com/Kays, L., 2011.

Mackenzie (2012). Achieving sustainable development: The Centrality and multiple facets of integrated decision-making. *Indian Journal of Global Legal Studies,* 247–285.

Major Zaragoza (2009). The Shared Meaning of Sustainability within the New Zealand Business Context and Its Implications. A Delphi Study. Unpublished Master of Business Management, University of Otago, Dunedin.

Merriam Encyclopedia (2001). Sustainable development: A critical review, Impact of social media on youth, 19(6), 607–621.

Paehlke (2001). Social work and sustainability: A council on Social work education, Global Commission.

Pempek *et al. (*2009). Sustainable development strategies: Tools for policy coherence. *Natural Resources Forum,* 136–145.

Peters (2012). Are we planning for sustainable development? An evaluation of 30 comprehensive plans, *Journal of the American Planning Association,* 66(1): 21–33.

Quan-Haase and Young (2010). What adolescents can tell us: Technology and the future of social work education. Social Work Education, 30, 830–846.

Redclift, M. (1992). The Meaning of Sustainable Development, Geoforum, 23(3): 395–403. Natural Resources Forum, 136–145.

Richard Mathews (2014). Sociology and environmental impact assessment. *Canadian Journal of Sociology/Cahiers Canadians de Sociologies* 22(4): 457–479.

Sheldon (2003). Human Development and Economic Sustainability. World Development, 28(12), 2029–2049.

Streeten, Burki, Haq, Hicks, and Steward (1981). The digital age and implications for social work practice. *Clinical Social Work Journal,* 40, 277–286.

Thogersen, S. (1997). Towards a 'science of sustainability': Improving the way ecological economics understands human well-being Ecological Economics, 23(3), 95–111.

Thomas Dyllick and Kai (2005). Enabling sustainable management through a new multidisciplinary concept of customer satisfaction. In: *European Journal of Marketing,* Vol. 39, 9/10, pp. 998–1012.

Will (2008). Exploring the ethics of social media. The New Social Worker Online. Retrieved 21/02/2013 from http://www.socialworker.com/home/Feature_Articles/Ethics/Must_I_Un-Friend_Facebook?_Exploring_the_Ethics_of_Social_Media.

World Commission on Environment and Development (WCED) (1987). Sustainable Urban Development Reader, Grooming cyber victims: The psychosocial effects of online exploitation for youth in Journal of School Violence 2(1): 5–18.

Role of Terrace Garden in Sustainability and Environment—A Case Study

Poornima Rao

Electrical Engineering Department, Fr. C. Rodrigues Institute of Technology,
Agnel Technical Complex, Navi Mumbai–400703, India
E-mail: punnag@yahoo.co.in

ABSTRACT: *As the world is heading towards the depletion of natural resources and the loss of forest/garden area due to urbanization, there is a dire need of terrace gardens. Due to the population explosion with a house for every citizen in the country, all the open areas are eaten away by concrete buildings. This has created the ecological imbalance, which can cause tremendous harm to our future generations. When we cannot avoid utilizing open spaces on the ground for the construction of buildings and other utilities, then at least the open spaces available above these buildings can be utilized for planta-tions and gardens to minimize the ecological imbalance, if not eliminate it altogether. There are many benefits of these terrace gardens, such as waste recycling, ecological benefits, energy conservation, water conservation, decorative enhancement of buildings, occupant's health benefits and attracting birds and insects. The list can be much longer. In this paper, a case study is considered highlighting the benefits of terrace garden and its significant role in sustainability and environment. The paper is based on the author's own experience in self-cultivated terrace garden.*

Keywords: Terrace Garden, Utilizing the Available Open Space, Sustainability, Energy Efficiency, Environment and Health.

1. INTRODUCTION

Due to the urbanization and growing population, there is a big threat to sustainability. There are many ways and means of maintaining this sustainability to some extent by the citizens. One of the measures would be to have a terrace garden to utilize the available open space in a productive way. It serves many purposes. But one important purpose definitely served would be for those who have a passion for gardening. Apart from that, it has many other benefits like ecological benefits, water conservation, energy conservation, decorative enhancement and attraction to birds and insects. Terrace gardens also contribute tremendously towards the health betterment of the occupants of the building. There are many people who are passionate of gardening but often are disappointed, as they may not be fortunate enough to have open spaces for the same. The author of this paper is one amongst them. Passion is important for a strong motivation to have a terrace garden, because it is not very easy to have and maintain a terrace garden as the one on the ground. There are many obstacles like the quality of construction, proper water-proofing of the terrace, strength of the building to withstand additional weight, the watering facility available, the drainage facility available and proper maintenance of the garden. Hence it calls for some investments and expenditure as well.

2. BENEFITS OF TERRACE GARDEN IN SUSTAINABILITY

2.1 Energy Conservation

Till the water-proofing level or membrane insulation level, both the conventional roof top and the one with the garden are similar from the heat transfer point of view. The elements of conventional roofs absorb the solar radiation during the day when the temperature of air is hotter than the roof and radiates this heat back to the surrounding air during the night when air temperature is less than that of roof top. Due this radiation of heat back to the environment from the roof tops, the urban areas experience warmer nights and hence there is a need for the coolers in the rooms below the roofs during the night, especially in the summer months.

Whereas the buildings with terrace gardens, this thermal profile changes since there is a direct shading of the roof [Green roofs-EPA] due to the garden and hence the roof top temperatures are always lower than the atmospheric temperature around and hence heat is continuously absorbed from the air around to keep the environment cooler. Also when the terrace gardens are watered in the late evenings, the roof tops are cooled further. Hence the room temperatures with the terrace gardens will be at least 3 to 4 [Scholars research library-2012] degrees lower than that of conventional roofs without gardens. This reduces the cooling load [Green roof –Wikipedia] on air conditioners in the summer, since the set temperature of thermostat can be 4 to 5 degrees higher or it may even eliminate the need for air conditioners altogether depending on the region.

2.2 Case Study

Comparison of Energy spent for air conditioners in the room without and room with terrace gardens is given below.

Energy consumption of air conditioner is based on several diverse factors [*BijliBachao* Team-2015] like area of the room, climatic conditions based on the region(cold, hot, humid), interior of room color-dark/light (wall, floor, curtains, furnishings), finishing-rough/smooth, number of people in the room (residential, office room, class room), type and color of light bulbs in the room (on/off), any other electrical/electronic gadget present in the room, number of solid objects in the room and the thermostat setting for temperature, because all these factors contribute for absorption of heat inside the room. Higher absorption of heat by these elements slowsdown the cooling process in the room by air conditioners and hence more energy needs to be drawn by the air conditioners.

Therefore, in our case study, all the above conditions are considered as moderate while comparing the energy consumed. The Table below shows the comparison for a temperature difference of 4 degrees in room temperatures only due to the terrace garden.

Table 1: Comparison of Energy Consumption in Room with/without Terrace Garden

Features for Comparison	*Room without Terrace Garden*	*Room with Terrace Garden*
The total area of the roof top considered for the case study	15 ft. × 30 ft.	15 ft. × 30 ft.
The recommended air conditioner size for the area of the room below	1.5 ton	1.5 ton
The type of air conditioner	Split	Split
Thermostat setting	20 degrees	24 degrees
Energy consumed for 1 hour (approx.)	1.2 units	1 unit (considered 4% lesser for every degree of temperature increase in thermostat setting)
Energy consumed for 6 hours	7.2 units	6 units
Energy consumed for 5 months in a year (being residential load)*	1080 units	900 units
Total units saved	180 units	

*Energy comparison is done for the room in the coastal region with a minimum usage of the air conditioner for 3 months in summer (March, April and May) and 2 months in post-monsoon period (October and November).

3. ROLE OF TERRACE GARDEN IN IMPROVING ENVIRONMENT

3.1 Improving the Air Quality

Presence of any greenery around always improves the air quality because, in addition to photosynthesis, they are able to absorb dust and other airborne toxins and during rainy seasons they wash them away into the drains. Hence the plants act as natural filters of dust and other toxins. Therefore plants clean the air in around the garden.

Fig. 1: Aerial Partial View of Roof Garden (l); Recycling of Wet Kitchen Waste (r)

3.2 Improving Occupants' Health

The organic yields obtained from one's own garden are free of chemical fertilizers and pesticides (if garden is maintained accordingly). The special herbs, vegetables and fruits which have medicinal values can also be grown and consumed to improve the occupant's health and hence improve their productivity.

3.3 Improves Ecological Balance

Terrace gardens contribute towards ecological balance by providing shelters for other inhabitants in nature like birds, earthworms and butterflies.

3.4 Easy Recycling of the Wet Waste

The wet waste generated in the kitchen can be easily recycled by pouring it in to the terrace garden. This way, cost of additional manure required for the garden can be saved. This helps in garbage disposal in a productive way.

4. CONCLUSION

Terrace gardens should be considered as a boon for maintaining sustainability and Environment as they provide Energy savings for air conditioners [Patric Dixon] apart from many other benefits. The diminishing open areas for private/public gardens in the cities are being compensated. The case study discussed indicates the energy savings of just one room in one house hold. When the same is applied for other rooms and buildings, the energy conservation is quite substantial and contributes towards reduction in carbon emission.

REFERENCES

BijliBachao Team–2015.
Green Roofs-United States EPA.
Green Roofs–Wikipedia.
Roof Garden Impact on Energy Savings–Patric Dixon.
Scholars Research Library–2012.
The author's own terrace garden as shown in the pictures.

A Study on Status of Logistics Infrastructure in India

Pratik Badgujar[1], Anuradha Bhagat[2] and Avdhesh Dalpati[3]

IPE Department, Shri GS Institute of Technology and Science, Indore
E-mail: [1]pratikbadgujar24@gmail.com; [2]anuradha.bhagat1607@gmail.com;
[3]adalpati@sgsits.ac.in

ABSTRACT: *The importance of Supply chain management was never realized well than present times. It is said that it is not the enterprises but their supply chains compete. With the advent of various transport options and information technology solutions this battle has turned only fiercer. Logistics is the crux of supply chain. It is logistics which ultimately decides whether a Supply chain will be able to deliver what it is aimed at. Logistics industry is mainly dependent on state of infrastructure in the country. With sound infrastructure only can the Logistics sector thrive and be productive. This paper aims to take a look at status of infrastructure available for Logistics management in India. A detailed analysis of the state of rail, road, air and waterways' infrastructure is done. The paper also compares India's progress in infrastructure with that of BRICS nations.*

1. INTRODUCTION

As per Council of Logistics Management, logistics is defined as "the process of planning implementing and controlling the effective and efficient flow and storage of goods, services and related information from the point of origin to the point of consumption for the purpose of conforming to customer requirements". Logistics is an important part of supply chain as any movement of materials forward or reverse and relevant information across supply chain is done by logistics.

Productive Logistics is critical today as more and more supply chains are turning global with global suppliers. Moser R. *et al.* (2011) also agree that supply chains faces challenges of logistics due to globalization. Without a good support from logistics adopting lean manufacturing cannot be viable. In the highly agile but complex environment of Just-in-Time and Kanban inventory management, the unpredictability of logistics performance is absolutely unacceptable both to companies and customers. So improving logistics services capabilities is imperative.

This paper attempts to make an account of current status of logistics infrastructure India and scope for its development. The paper takes a look on the current infrastructure status and future requirements to further boost logistics services in the country.

2. LITERATURE SURVEY

Logistics has been an underutilized and under considered industry so far. Logistics matters much more for the overall health and dynamism of the world economy than policy-makers appear to realize.

According to World economic forum (2013) if all countries improve their logistics performance while simultaneously reducing supply chain barriers to half the level observed in best-performing country in their respective regions, global GDP could increase by 2.6%.

The ratio of trade to GDP of world as a whole has increased by 20% over the century (Figure 1). The growth in trade results in parallel increase in demand for superior logistics services (World Economic Forum, 2013). Hence as the Indian economy grows in future, the logistics industry is also expected to expand.

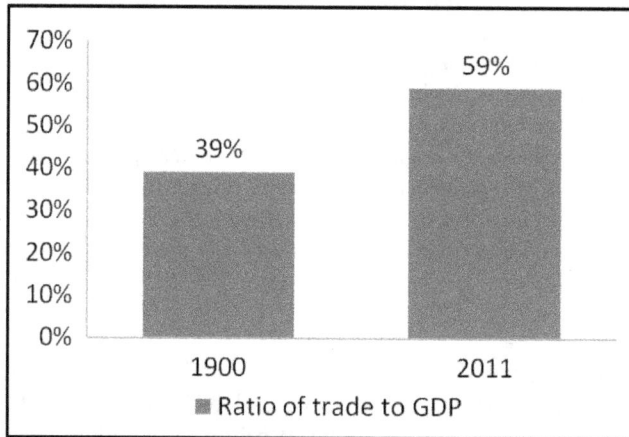

Fig. 1: Ratio of Trade to GDP of World as a Whole (WEF, 2013)

2.1 Indian Scenario

The logistics market in India was valued at 5.6 trillion in 2010 which is expected to reach 17 trillion by 2015 with a Compounded Annual Growth Rate (CAGR) of over 8% (Price water house coopers [PWC], 2010). A major part of transport in India is done through roads which can be inferred from the table below:

Table 1: Share of Transport Sector in Overall GDP of India (in % age)
(Ministry of Railways, [MOR], 2015)

	2008–09	2009–10	2010–11	2011–12	2012–13
Overall Transport *of which*	6.6	6.6	6.5	6.6	6.7
Road Transport	4.7	4.7	4.6	4.8	4.9
Railways	1.0	1.0	1.0	1.0	0.9
Air Transport	0.2	0.2	0.3	0.3	0.3
Water Transport	0.2	0.2	0.2	0.2	0.2

Indian industry spends a high 14% of its GDP on logistics which is higher as compared to 10-11% in Europe and 9% in USA (Boeing Commercial Airplanes, 2013), (Bharat Thakkar, 2013). Market constraints such as highly fragmented structures, poor infrastructure and complicated tax structure are primarily responsible for such higher logistics costs (Singh and Gandhi, 2011). The top-10 listed logistics service providers in India have only about 2 per cent share in the overall logistics industry (India Brand Equity Foundation [IBEF] 2013).

Another significant difference in the demographical situation of India and other developed nations is that in India the retail is mostly unorganized. Retail Density *i.e.* number of retail outlets per square km of distance is very high with about 12 crore retailers spread all over the country (Ketkarand Vaidya, 2012). Most of the population resides in villages which are spread all over the length and breadth of country due to which local help is inevitable compelling any supply chain to have multiple parties involved to reach outlets. All this makes logistics a difficult job.

(Sunil Chopra *et al.,* 2010) also identified unique structure of Indian supply chain which according to them has developed into its peculiar format. They further add that in India manufacturers do not enjoy substantial control over supply chain which makes managing supply chain partners difficult. Due to such uncertainties and under developed environment, supply chains in India end up to be multi party and locally optimized systems having multiple controlling parties. All these distinct characteristics make improvements in logistics sector a challenging task.

3. LOGISTICS INFRASTRUCTURE IN INDIA

3.1 Road Infrastructure

Road infrastructure is used to transport over 60 per cent of total goods and hence roads are backbones of Indian logistics sector. According to the Ministry of Road Transport and Highways, road freight in India is expected to reach 1,835 Billion Ton Kilometers (BTKMs) by 2016–17 (IBEF 2013).

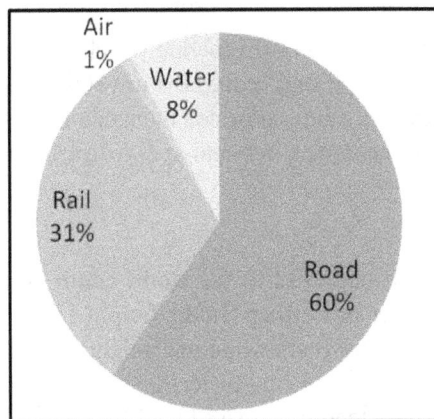

Fig. 2: Share of Different Modes in Indian Logistics Industry
(National Skill Development Corporation [NSDC], 2014)

India has total of 48, 65,394 km of roads of which only 26, 98,590 km (which is 55.46%) are surfaced (Ministry of Statistics and Programme Implementation [MSPI] 2014). The road density per 1000 persons is 2.3 km in urban areas. Rural areas are lagging behind in accessibility to roads whereas urban roads are more congested with only 1.27 km. of road per 1000 persons. Percentage of expressways or highways in India is also very less at 21,181 km out of 46,89,842 km (about 0.45%), as compared to China's 2.27% and USA's 3.98% (Wikipedia, 2014).

The present network of National Highways is about 79,243 km which serve as the main road network of the country. Even though Expressways and National Highways constitute only about 1.7% of the length of all roads about 40% of the road traffic is carried by them.(Ministry of Rural Development, 2011).

The anomaly is that increase in number of motorized vehicles is not supported by parallel increase in road lengths in country. While the motor vehicle population has grown from 105 million in 2007–08 to over 159 million in 2011–12 with Compounded Annual Growth Rate (CAGR) of 11%, the road network has expanded from approximately 4.2 million km in 2007–08 to 4.8 million km in 2011–12, at a CAGR of 3.4% only (MSPI, 2014).

Although the picture seems to be gloomy there is silver line there. The government is investing in infrastructure to make the sector more competitive, efficient and cost-effective. Changing regulatory set-up, government incentives, infrastructure development projects, increasing transparency, relaxed tax structures to encourage participation of foreign and private players in the sector and dedicated logistics parks and Free Trade Warehousing Zones (FTWZ) are expected to speed up growth of Indian logistics sector as well as contribute positively towards its identification as an independent sector by the government (Singh and Gandhi, 2011).

Road Projects such as north-south and east-west road corridors, the Golden Quadrilateral and other NHAI projects are promoted to improve connectivity in the country (MSPI 2014). National Highway Development Program (NHDP) which is the nodal agency for development of Highways and expressways has established robust objectives.

Apart from these ten expressways are in the pipeline which once completed will increase the speed of travel amongst the cities and thus reduce the travel time. Such extensive construction and expansion of the road infrastructure is expected to enable the construction of warehouses at common points of collection to become a hub-and-spoke distribution system.

3.2 Rail Infrastructure

India has one of the largest rail networks in the world and it is also the cheapest means of transport in the country. It runs more than 7,000 freight trains per day carrying about 3 million tons of freight. In the last 64 years while the freight loading has grown by 1344% and passenger kilometers by 1642%, the route kilometers have grown only by 23%. The percentage of double way and multiple route length increased by 289% which is also less as compared to increase in demand.

Table 2: Change in Freight Traffic over the Years

Item	1950–51	2013–14	% Change
Route length	53,596	65,806	+23%
Freight carried (million tonnes)	73	1,054	+1344%

About 40.36% of rail sections are running at 100% or above line capacity. In recent period from 2011 to 2015 new lines have registered a growth of 74%, while doubling and electrification have risen by 167% and 21% respectively.

The rail connectivity in some parts of the country is highly restricted. For example, the rail connectivity in North eastern states is negligible with all seven states has a meager 168 km of rail network available (IBEF, 2013).

The government is focusing on accelerating the development of the network for freight terminals through private investments. The government has permitted 15 private train container operators to run container trains. Various schemes such as Special Freight Train Operator (SFTO) and Private Freight Terminals (PFT) programs are introduced to encourage investments in rail racks and terminals. Successful implementation of these schemes is expected to improve logistics efficiency in India (Singh and Gandhi, 2011). Private sector is also involved in rail transportation, but has a minor role. The government has involved many private players in the operation of container trains also (PwC, 2010) which will lead to increase in competition resulting in improvement in performance.

Dedicated Freight Corridors (DFCs), Diamond Rail Corridor Project are likely to improve inter conectivity of metro cities as well as Indian Railways' freight carrying capacity. Improvement of rail network in eastern corridor is also imperative to facilitate the development of the proposed Trans–Asian Railway involving infrastructure investments in India, Bangladesh, and further eastern countries (Singh and Gandhi, 2011). Post completion of the DFC project in 2016–17, Dedicated Freight Corridor Corporation of India Limited (DFCCIL) expects DFCs to carry 30% of total freight traffic in India till 2021–22 (MSPI, 2014).

3.3 Challenges before Railways

These are some major challenges before Indian Railways which have to be solved to improve its logistics performance and increase its share in Indian logistics industry (Ministry of Railways [MOR], 2015):

- Absence of liquidity,
- Law and Order problems,
- Delayed projects
- Inclusion of technology
- Improving timeliness and safety of freight.

Airway Transport

Worldwide air freight is expected to more than double till 2033, increasing from 200 billion Revenue Tonne Kilometers (RTKs) in 2013 to 512.3 billion RTKs. Air cargo volume in India grew at a CAGR of about 8.5 per cent from 1998–99 to 2012–13 (Ministry of Statistics and

Programme Implementation, 2014).The growth rate of domestic Indian air cargo market was 6.9% annually. In 2013, domestic Indian air cargo increased 2.3% over 2012 to 3, 71,000 tons. The expansion of Indian air cargo industry is projected to continue at a rate of 6.3% per year from 2013 to 2033 (Boeing Commercial Airplanes, 2013).

Table 3: Growth Statistics of Indian Air Cargo Industry

Duration	Compounded Annual Growth rate (CAGR)	Volume	Reference
1995–96 to in 2010–11 (Total)	8.7%	0.68 million tons to 2.39 million tons	Acaai pp. 8
1995–96 to 2010–11 (Domestic cargo growth)	9.7%	0.22 million tons to 0.89 million tons	Acaai pp. 8
1995–96 to 2010–11 (International cargo growth in India)	8.2%	0.46 million tonnes to 1.5 million tonnes	Acaai pp. 8
From 1998–99 to 2012–13 (Air cargo)	8.5%		IBEF, 2013
Forecast			
2013–2031 (total)	11.2%	NA	Acaai pp. 8
2013–2031 (domestic)	10.4%	NA	Acaai pp. 8
2013–2031 (international)	11.7%	NA	Acaai pp. 8
Till 2018–19	NA	5 million tons	MunmunBasak *et al.,* 2013
2031 (total)	NA	18.19 million tons	Ministry of Civil Aviation, 2012
2031 (domestic)	NA	6.9 million tons	Ministry of Civil Aviation, 2012
2031 (international)	NA	11.29 million tons	Ministry of Civil Aviation, 2012

The International cargo throughput is expected to grow 7.5 times (CAGR of 11.7%) till 2030–2031. During the same period total cargo throughput at Indian airports is expected to grow 7.6 times at CAGR of 11.2 percent and domestic cargo throughput is expected to grow 7.8 times at CAGR of 10.4 per cent. (Boeing Commercial Airplanes, 2013).

Intra-Asia is likely tobe the third largest air cargo market by 2033. The share of world air cargo traffic associated with Asia is expected to increase from 51.3% in 2013 to 61.1% in 2033 (Boeing Commercial Airplanes 2013).

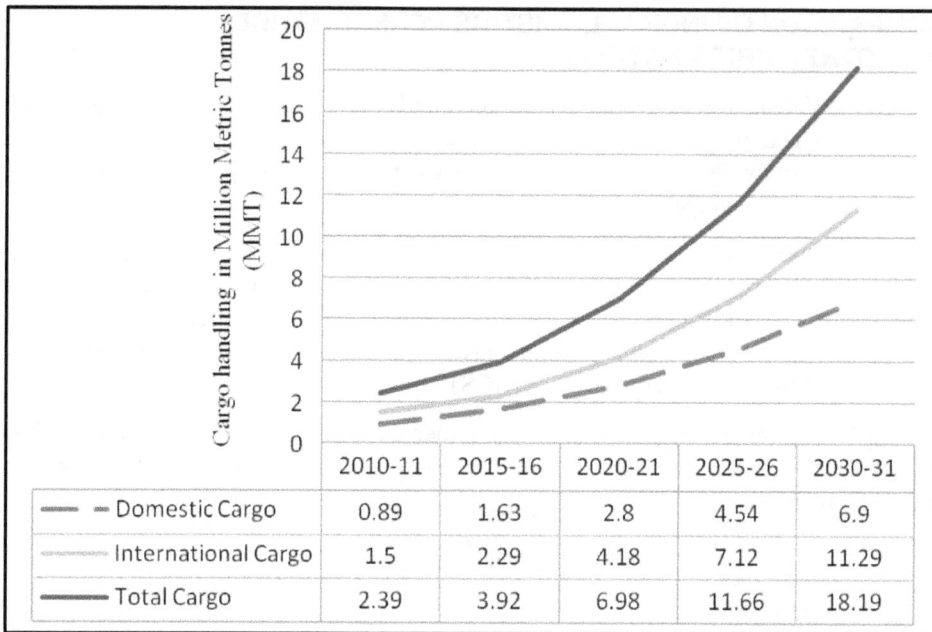

	2010-11	2015-16	2020-21	2025-26	2030-31
— — Domestic Cargo	0.89	1.63	2.8	4.54	6.9
International Cargo	1.5	2.29	4.18	7.12	11.29
Total Cargo	2.39	3.92	6.98	11.66	18.19

Fig. 3: Projection of Air Cargo Traffic (Ministry of Civil Aviation, 2014)

3.4 Water Transport

It is a well acknowledged fact that water-borne transport is much safer, cheaper and cleaner compared to other modes of transportation. As stated above present logistics costs in India are quite higher than many other countries and one of the reasons is its underutilized coastline. India has a vast coastline of over 7,000 km. Although it has more than 200 ports (comprising 12 major and more than 180 small ports) the Indian share of water transport in international trade is less (Ministry of Shipping, 2014).

The government is considering revival of India's coastal lines by project 'Sagarmala'. The project 'Sagarmala' aims at developing existing ports to world standards. Development of ports are likely to integrate the hinterland projects of industrial and freight corridors with the maritime developments to offer efficient and seamless transport for both domestic and international sectors thereby reducing logistics costs.

The government is exploring options of developing inland waterways in the country. Although the number of rivers having round the year flow is less, developing waterways on available routes will certainly add to logistics capacity of the country. Shipping routes through coasts and inland waterways are primarily used for transportation of bulk freight. India possesses about 14,400 km of inland waterways. Over 3,600 km are navigable by large vessels, 55 per cent of which is being used. To utilize full potential of waterways, six national waterways have been declared. (IBEF, 2013).

With all the above measures falling in place the Indian water freight capacity will increase and will also help to reduce logistics costs in the country.

4. COMPARISON OF INDIA'S LOGISTICS PERFORMANCE WITH OTHER BRICS NATIONS

World Bank is calculating Logistics Performance Index (LPI) since 2007. The index includes more than 150 countries. LPI and its indicators have been constructed from information gathered in a worldwide survey of the companies involved in freight movement acrossthe world (Connecting to compete, 2007).

LPI compares nations on six parameters *viz.* Customs, Tracking and tracing, Infrastructure, International shipments, Timeliness and Logistics quality and competence. Domestic logistics cost was included in LPI index of 2007 but was dropped from LPI calculations in successive reports. Table 1 shows a comparison of rankings of vibrantly developing nations Brazil, Russia, India, China and South Africa (acronym BRICS).

Having an LPI lower by one point (say, 2.5 rather than 3.5) implies six additional days for getting imports from the port to a firm's warehouse and three additional days for exports. (Connecting to compete, 2007).

Table 4: LPI Ranking of BRICS Nations

Country	*2007*	*2010*	*2012*	*2014*
Brazil	61	41	45	65
Russia	99	94	95	90
India	39	47	46	54
China	30	27	26	28
South Africa	24	28	23	34

Source: Connecting to Compete, 2007, 2010, 2012 and 2014.

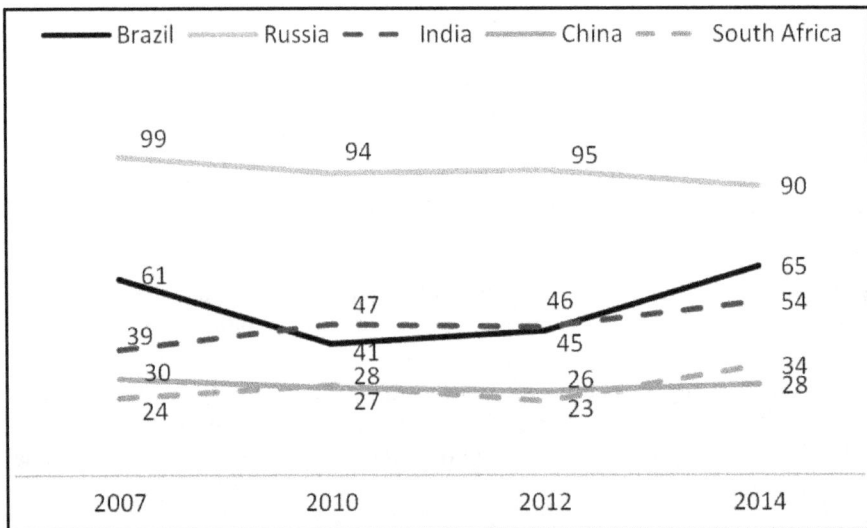

Fig. 4: LPI Rankings of BRICS Nations
Source: Connecting to Compete, 2007, 2010, 2012 and 2014.

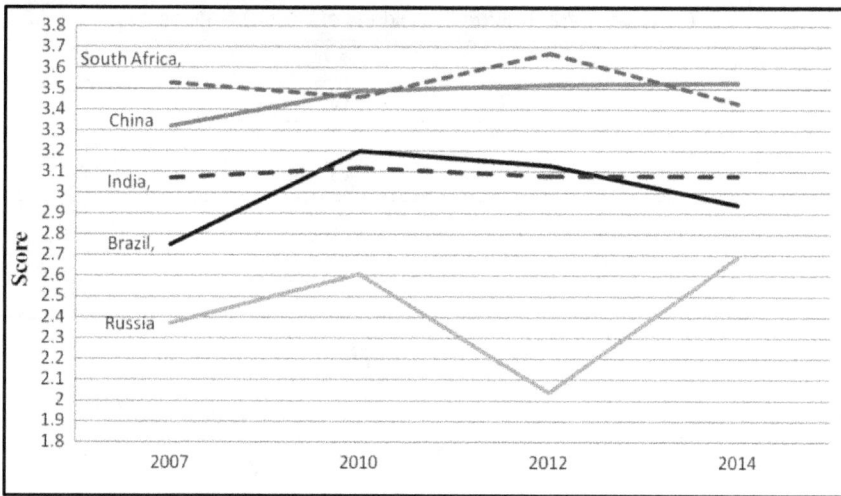

Fig. 5: LPI Scores of BRICS Nations

Source: Connecting to Compete, 2007, 2010, 2012 and 2014.

A comparison of LPI rankings of BRICS nation indicates that India's performance on this index has deteriorated over the last 7 years as it slipped to 54th spot in 2014 from 39th in 2007. The performance of Russian Federation is poorest and that of China has been the best among BRICS nations.

Taking a look at the infrastructure rankings of India in LPI reveals more about the past infrastructure developments in India. A comparison of India's ranking is also done with that of BRICS nations.

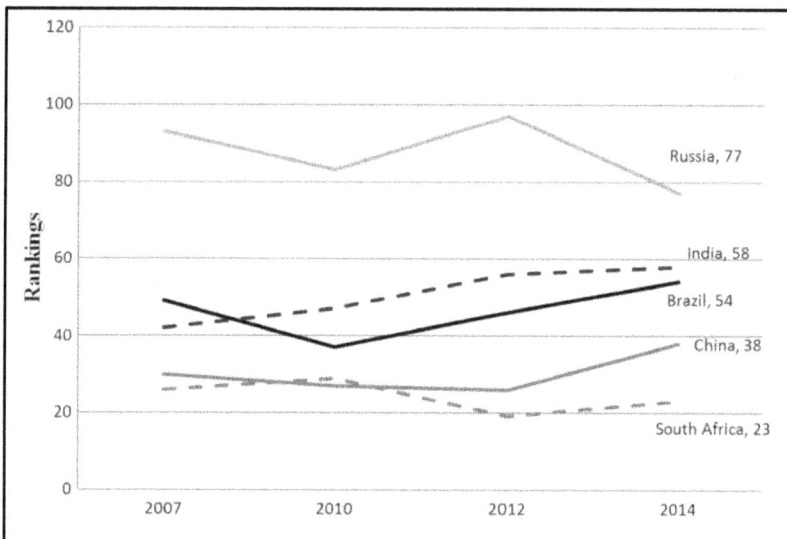

Fig. 6: LPI Infrastructure Ranks of BRICS Nations

Source: Connecting to Compete, 2007, 2010, 2012 and 2014.

Table 5: 'Infrastructure' Rankings of BRICS Nations

Country	2007	2010	2012	2014
Brazil	49	37	46	54
Russia	93	83	97	77
India	42	47	56	58
China	30	27	26	38
South Africa	26	29	19	23

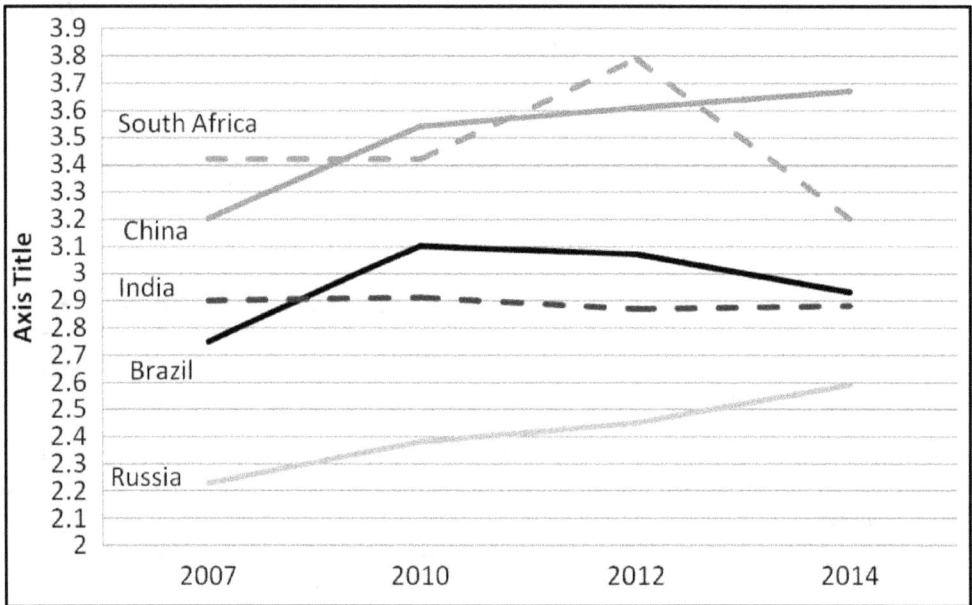

Fig. 7: LPI Infrastructure Scores of BRICS Nations

Source: Connecting to Compete, 2007, 2010, 2012 and 2014.

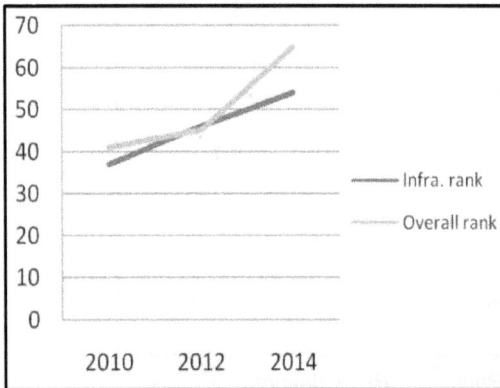

Fig. 8 (a): Brazil

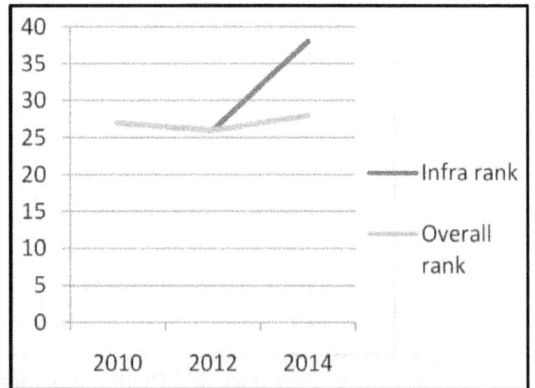

Fig. 8(b): China

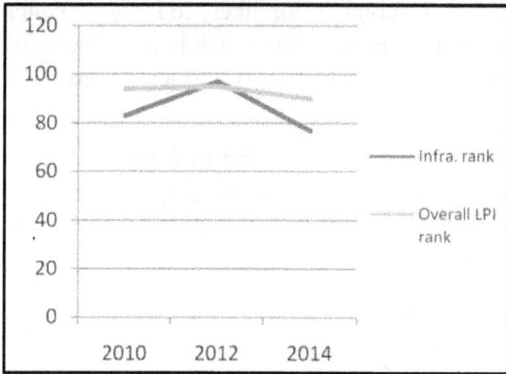

Fig. 8(c): Russia

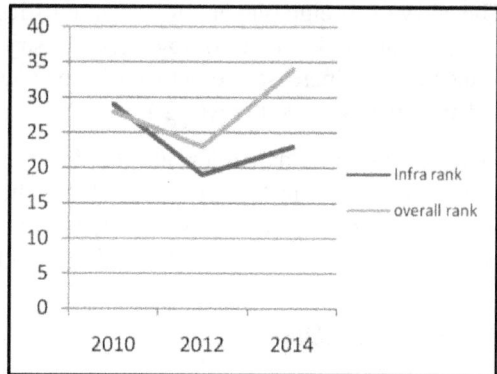

Fig. 8(d): South Africa

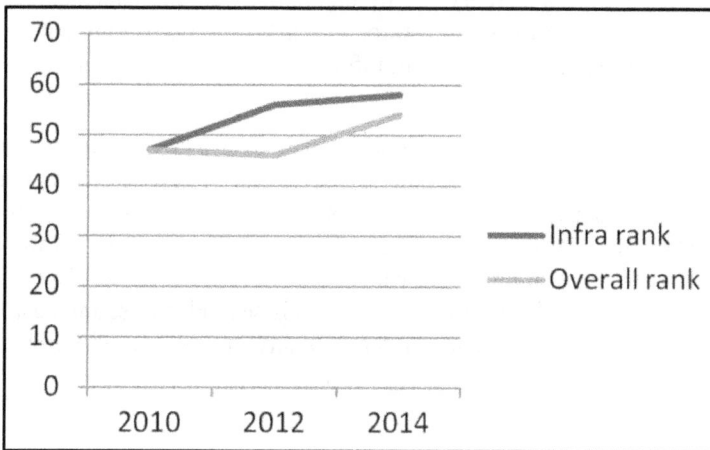

Fig. 8(e): India

Fig. 8: A Comparison of Overall Ranking of BRICS Nations with their Infrastructure Rankings

On comparing overall ranks of BRICS nations with their infrastructure ranking interesting conclusions are drawn. The comparison reveals that for Brazil, Russia, China and South Africa trend of overall rank is similar to that of respective infrastructure development. However in case of India same doesn't hold true. The change in India's 'overall' rank is not in line with that of Infrastructure.

5. FUTURE TREND

Privatization of transport and logistics operations is likely to play important role in India in the coming times (PwC, 2010). The government has eliminated foreign direct investment caps for the Indian shipping industry. Increasingly, investor friendly frameworks and outstanding growth opportunities in the Indian logistics industry are attracting many multinational logistics companies (PwC, 2010).

The entry of foreign players in Indian logistics sector the competition likely to increase which will promote high growth of logistics industry in coming years. Also growth in manufacturing industries will increase the demand in logistics industry and will lead to further expansion of Indian logistics market (PwC, 2010).

Changing business dynamics and the entry of global 3PLs have also led to many significant structural changes in Indian supply chains including logistics and warehousing services. The Indian logistics sector is now evolving from an existential state to a proactive contributor to efficient and effective supply chains (Singh and Gandhi, 2011).

6. CONCLUSION

The length of road and rail network has increased over the years and still there is ample scope for further development. The government has come up with several initiatives to improve logistics infrastructure in the country. With this initiatives implemented successfully, the logistics sector is expected to thrive in future. The integrated development of road, rail, air, sea and Inland water transport and air freight infrastructure will be helpful to streamline logistic operations.

The Indian Logistics sector has shown good growth in previous years but it has failed to outweigh that of other BRICS nations. India's progress on infrastructure front has rather been tardy over the years. As per the latest rankings of 2014 India ranks 4th amongst BRICS nations only better than Russian federation. In previous years the growth of Logistics infrastructure in India has been moderate which is evident from its consistent LPI infrastructure scores. The improvement in Infrastructure scores is second lowest amongst BRICS nations outweighing South Africa only. However the infrastructure score of South Africa is quite higher than that of India.

REFERENCES

Bharat Thakkar (2013). 'Two years were either too long of too short', ACAAI, Vol. 4, Issue 3, July–Sept. Issue10, pp. 3.

"Building the Supply Chain of the future" McKinsey (2011). Available at: www.mckinsey.com/insights/operations/building_the_supply_chain_of_the_future. (Accessed on Dec. 23, 2013).

Chopra Sunil, Meindl Peter, and Kalra D.V. (2010) 'Supply Chain Management: Strategy, Planning and Operations', India. pp. 154, 246, 307, 417.

'Connecting to compete' (2007) World Bank, Washington DC, USA.

'Connecting to compete' (2010) World Bank, Washington DC, USA.

'Connecting to compete' (2012) World Bank, Washington DC, USA.

'Connecting to compete' (2014) World Bank, Washington DC, USA.

'Enabling Trade: Valuing Growth opportunities' (2013). World Economic Forum, New York, USA.

'Human Resource and Skill Requirements in the Logistics, Transportation and Warehousing Sector (2013–17, 2017–22) – Draft Version' (2014). NSDC, New Delhi, India.

'Indian Logistics industry: Gaining momentum' (2013). Indian Brand Equity Foundation, Gurgaon, India.

Kaushika Madhavan, Saurin Doshi, Nithin Chandra and Manish Pansari (2013). A.T. Kearny study for CSCMP. 'Creating competitive advantage through the Supply chain: Insights on India', pp. 8, 12, Mumbai, India.

Ketkar Manisha and Vaidya O.S. (2012). 'Study of emerging issues in Supply risk management in India', *International conference of emerging economics- Prospects and Challenges, Procedia– Social and Behavioral Sciences.* Issue 37, pp. 57–66.

Khare Arpita (2008). 'Supply Chain Collaborations changing the face of Indian Automobile', *The International Journal of Applied Management and Technology*, Vol. 6, Num 1.

Madhavan. K., Doshi. S., Chandra. N. and Pansari. M. (2013). 'Creating Competitive advantage through the Supply chain: Insights in India', *A.T. Kearny Analysis* Mumbai, India.

Ministry of Civil Aviation (2012). 'Air Cargo Logistics in India, A working group report', New Delhi, India.

Ministry of Railways (2015). "Indian Railways: Lifeline of the nation (A White Paper)", Government of India, New Delhi.

Ministry of Rural Development (2011). 'Working Group on Rural roads in 12th five year plan' Government of India, New Delhi.

Ministry of Shipping (2014). 'Concept Note on Sagar Mala Project: Working Paper', New Delhi, India.

Ministry of Statistics and Programme Implementation (2014). 'Infrastructure stastics–2014', Vol. 1, Issue 3, New Delhi, India.

Moser, R., Kern, D., Wohlfarth, S. and Hartmann, E. (2011). "Supply Network Configuration benchmarking – Framework development and application in the India automotive industry", '*Benchmarking: An International Journal*', Vol. 18, No. 6, pp 783–801.

Munmun Basak, Martin west and S.P.S. Narang (2013) 'Forecasting air cargo demand in India', International Journal of engineering sciences and innovative technology, Vol. 2, Issue 6, pp. 398.

National Skill Development Corporation (2014), 'Human Resource and Skill Requirements in the Logistics, Transportation and Warehousing Sector (2013-17, 2017–22) –Draft Version', New Delhi, India.

'Only India in BRIC nations has slipped on World Bank's Logistics Performance Index', *The Hindu Business Line.*, New Delhi., 2012.

Retrieved from http://en.wikipedia.org/wiki/List of countries by road network size, (accessed on 27/02/2014 at IST 22:11)

Singh Rachna Nath and Gandhi Dilraj (2011) 'Building warehouse competitiveness in India', Price Waterhouse Coopers, New Delhi, India.

Srivastava, S.K. (2006), "Logistics and Supply Chain Management Practices in India", 6th Global Conference on Business and Economics.

'Transport and logistics 2030' (2010) Price Waterhouse Coopers, New York, USA.

'World Air cargo Forecast 2014–15.' Boeing Commercial Airplanes (2013), Seattle, Washington USA.

Economic Growth and Environmental Stability at Equilibrium Point: A Sustainable Business Model for the World

Ravikumar Gajbiye[1] and Kavita Laghate[2]

Jamnalal Bajaj Institute of Management Studies, University of Mumbai,
Churchgate, Mumbai–400020, Maharashtra, India
E-mail: [1]ravikumargajbiye14@jbims.edu; [2]kavitalaghate@jbims.edu; director@jbims.edu

ABSTRACT: *Globally, there is a paradigm shift of businesses from profitability towards Sustainable growth. Sustainability is defined as the quality of protecting the environment by preserving natural resources, thereby supporting long-term ecological balance. Ecological-Economic Models describe the effect of the ecosystem landscape on business. Sustainability involves supporting long-term ecological balance. Economic development, social development, and environmental protection are the three pillars of sustainable growth. Natural resources that fuel economic growth are non-renewable and scarce, and are being rapidly depleted by nations to gain power through economic supremacy.*

Ecological Economic Models describe the effect of the ecosystem landscape on business. Equitable Growth can be achieved through constant UGDP modeling. Sustainable Business Models require maintaining equilibrium between economic growth and ecosystem stability. The cumulative valuation of natural resources in RGDP units which when equated with UGDP gives an ideal equilibrium point for Economic-Environmental balance. It is necessary to maintain responsible stewardship of the planet through the role of a global ombudsman.

Keywords: Economy, Environment, Sustainability, Equilibrium, UGDP, RGDP, Ombudsman.

1. INTRODUCTION

As resurgent winds of change blow in favor of a revitalized Indian economy under an able and stable leadership, the benefits of a robust financial system and sturdy commercial infrastructure are being sighted on economic horizons. Challenges abound in environmental and social domain, which call for sustainable development. But sustainability cannot exist in isolation; India needs to have an ambitious global vision; thinking beyond participatory and catalytic roles, towards global leadership.

Sustainability does not mean consistently deriving economic profits; it means keeping the world resources intact for future generations while simultaneously fulfilling the unsatisfied needs and wants of present human society. Sustainable Business Models based on triple bottom line principle, cater to business, government and society. Their strategy lies in creating minimal negative impact on society and environment while earning profits.

Strategic Sustainability initiatives for most business models involve accuracy and reliability in reporting, efficiency in business processes, and marketing proficiency while striving to reduce environmental footprint. The "take-make-waste" economy thrives on the draconian theory that "the primary aim of business is profit-making".

The Current relationship between a business organization and its related constituents is as depicted in the following Figure 1.

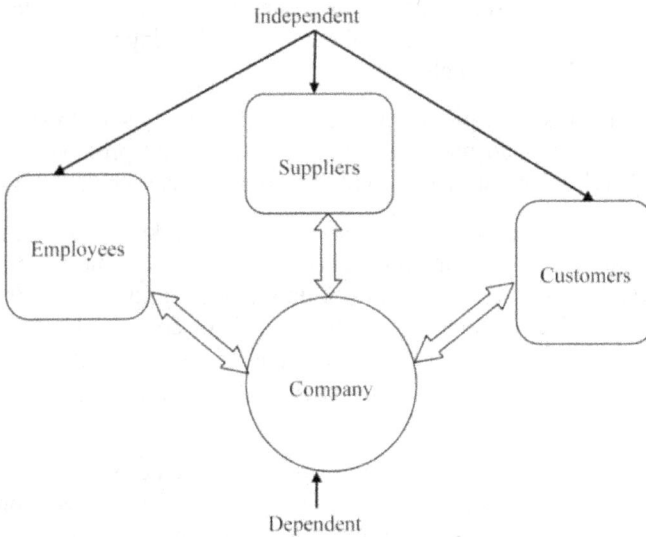

Fig. 1: No Causal Direction. Arrows Show Mutual Relationship

Exploitation of resources have created immense economic wealth for the world, but the beneficiaries of these spoils are only few. World resources are being destroyed to fuel the greed of money-hungry power-mongers. Economic wealth is the source of power: but this is passé. Large Centralized Systems are obliterating smaller decentralized energy systems. New Power axis: Energy systems working in an energy marketplace.

The importance of IT-enabled businesses in the new world order cannot be understated. Technology platforms can redefine businesses. Social networking sites have connected people from the most common denominator of nations spread all over the globe, unifying them into one universal brotherhood of mankind unified by cyberspace. Awareness of environmental facts will push the public demand for environmentally sustainable goods and services. No distinctions now remain, and knowledge is become freed form the domain of the classifieds. People are being more aware of their universe, and increasingly resentful of the exploitation of the natural resources for gainful wealth of a few prosperous nations.

2. TRENDING PATTERNS IN SUSTAINABLE BUSINESS MODELS

Sustainable Entrepreneurship patterns involve dematerialization, alternative market place, and social capital. The current trends that drive businesses are environmental occurrences like changing regulations, rising non-renewable fuel prices, falling prices of renewables, improved energy storage solutions, and decentralized energy options. Energy Businesses are transforming their genome from "Power Seller" to "Renewable Energy Enablers". They are adding "Social Value" by using the product as a service model, and are transforming from Volume to Value business. Businesses are embedding sustainability into their underlying business structure.

Businesses are facing the substantial challenge of transitioning to a low-carbon economy constrained by dwindling natural resources, while increasing consumption rates push demand on the limited supply of non-renewable raw materials. Giant integrated businesses with centralized planning and international consolidation are challenged by competition, political initiatives, regulation and structural changes.

Innovative business models offer pioneering companies an early start toward the future. They can signal to consumers how to make sustainable choices and provide reward for both the consumer and the shareholder. Innovation will ensure business success in a reset world.

It is necessary to integrate the cost of environmental markets into consumer goods by commoditizing the value of environmental services into business models to help consumers make smart choices. Combining the benefits of de-materialization with a disruptive business model prevents increased environmental destruction from continued over-exploitation, and works toward sustainability. Customer-Supplier relationships can be used an effective lever to support sustainability for encouraging viable businesses that provide either explicit social or environmental benefits.

Society today is fed on the concept of over-consumption, which drives businesses to increasing profits, but the challenge now is to make consumers content with less. Companies that can adapt and innovate themselves through economic or environmental crises, emerge stronger by innovating their business models with sustainability.

Business need to reinvent and innovate in technologies, products, services and business models to fundamentally change the value proposition they offer. Disintermediation within the industry will drive them beyond the traditional vertically-integrated utility model.

Businesses today are driven by the "Circular Economy Concept" and "Sharing Economy Business". This involves generating more value to customers without using more resources. A restructuring of business models is possible only for smaller companies or for start-ups who have to start from scratch, with innovativeness ingrained into their foundation structure. For market leaders with global presence, restructuring of the underlying business framework and to transform entire business models for sustainability purpose, would be a retroactive exercise.

3. LITERATURE REVIEW

Natural resources from undeveloped nations make the world economy grow, but do not contribute to their own growth. Rich nations use energy faster and are less sensitive to GDP growth (*Rögnvaldur Hannesson*[1]). Non-renewable energy consumption has a more significant effect on economic growth (*Usama Al-mulali*[2]). *Robert Repetto*[3] states that current System of National Accounting undermines the idea of sustainable development promoted by the WECD. National income accounting systems used by governments to assess macroeconomic performance tend to divide countries between developed and developing.

Ricardo Aguado and Jabier Martinez[4] highlight the inconvenience and unsuitability of using GDP as a proper indicator for economic sustainability, while identifying the main features of a proper measure of sustainability.

William Lyakurwa[5] stresses the need to integrate development, climate change and environmental sustainability. *Emil Urhammer*[6] states that economic growth is a powerful agent that determines policies regarding issues such as climate change, loss of biodiversity and pollution. *CalistaRajasingham*[7] writes about holistic and alternate development, poverty measures, and unemployment and inequality leading to Third World Exploitation.

Al-Rawashdeh et al.,[8] suggest that a strict policy Formulation is a necessity for environmental degradation. *Charles Perrings*[9] writes about change analysis in economy-environment systems, with emergent collaboration between ecology and economics. There is a need for including natural resources in measuring economic growth and a re-look at replacing capitalist structure based on greed with a more globally inclusive outlook, (*Bart Hobijn and Charles Steindel*[10]).

MoscosoBoedo, Hernan J.[11] considers macroeconomic equilibrium at universal level too statistical for a larger scope of demography or ethnicity. Alternative economics to find ways in which economies are made more ecologically benign and humane is advocated by *Laszlo Zsolnai.*[12] *Breuer, Dave* [13] considers environmental disaster as economic progress, and the rich-poor gap equity factor.

3.1 Gaps Analyzed in Literature

A universal model is yet to be established which is applicable for all nations of the world, covering all economic systems through lasting periods of times. The research advocates a sustainability-based business model with a global spread based on inclusion of natural resources with accurate and reliable valuation mechanism into mainstream world economics for equitable growth.

Sustainability entrepreneurship in ingrained into the business mechanism by the pragmatic utilization of natural resources for a regulated economic growth. The excesses of revenues must be ploughed back into the common revenue pool, and shortfalls must be adjusted by withdrawal from the common pool. The ultimate aim is to distribute equally among all nations of the world the economic produce of the natural resources that are taken from the earth for economic activity.

4. THEORETICAL FRAMEWORK

Ecological-Economic Models explain valuation of ecosystems by improving methods for ecosystem valuation. An interrelated model must describe the effect of the ecosystem landscape on the quality and value of goods and services and therefore, on human decisions under different regulatory regimes and alternative regulations. Economic models which overlook the technical interdependence of ecosystems have been surpassed by Ecological models which contain more quantitative parameters. John Tschirhart[14] favors inclusion of micro-behavior in creating adaptive ecological systems using economic optimization techniques without additional ecological detail into economic models.

4.1 Data and Sampling

Sampling: From the sample universe of 179 nations of the world, 12 nations are chosen as sampling units based on convenience and judgment sampling method. The economic and environment related data is collected from secondary source for a time period of 2009 to 2013. This is pooled cross-sectional data. Data of Land area, Population and Agriculture contribution to GDP is used. Six basic minerals are considered based on utility of natural resources towards Economic Growth.

Economic factors include GDP, Consumer Expenditure, Government Expenditure and Capital Investment made by Industry. It is a Pooled cross-sectional and Time series data (Panel data), with Periodicity: - Annual

Sample Size: The samples are chosen as tabulated below.

Table 1: List of 12 Sample Nations

Sample Nations	
Australia	New Zealand
Brazil	Russia
Canada	South Africa
China	Spain
Germany	UK
India	USA

Table 2: List of 6 Sample Minerals

Commodity	Weight
Coal	Metric Tons
Silver	Kg
Gold	Kg
Iron ore	Metric Tons
Petroleum	Metric Tons
Uranium	Tons

Sampling Database Used for Formulating Models

For RGDP Model: Database for Six Minerals important for Economic Growth

Table 3: Variable Type

Attribute	Variable Factor
Energy Source	Coal, Petroleum
Infrastructure Development	Steel
Economic Value	Silver, Gold
Scientific and Technological Growth	Uranium

Data Source: Data is collected from Secondary sources, historical and current.
Data Type Based on Variable Characteristics: The typology of data is given below.

Table 4: Variables, Constructs, Characteristics

Dependent	Independent	Moderating	Intervening
GDP (Y)	Population (X)	Exports (M)	Government Spending (G)
UGDP	Land (L)	Imports (N)	Consumer Spending (C)
RGDP	Agriculture-Revenues (Z1)	Investments (I)	
	Natural Resources (Z2)		

Characteristics of the Data for statistical purpose are as follows:

Table 5: Statistical Data Characteristics

Type of Data	Pooled Cross-Sectional and Time Series Data (Panel Data)
Periodicity	Annual
Unit of Analysis	Nations (the level of aggregation of data)
Study Setting	Contrived, Artificial, Stimulated
Time Horizon	One shot/cross-sectional
Variables (Objects)	Economic Growth of Nations of the World
Attributes/Characteristics	GDP (Y), Investment (I), Consumer Expenditure (C), Government Spending (G), Exports (X), Imports (M)

Aggregation of Data: Data is aggregated for statistical analysis purpose as follows:

Table 6: Statistical Data Aggregates

Over Different Time Periods	Time series data
Group of Nations in One Year	Cross-section data
Group of Nations Over many Years	Panel data
Multivariate Data over Several Geographic Areas	Aggregate cross section data
developed, Developing, Transient Economies	Strata of nations
Multivariate Data over Time for all Universe	Aggregate time series data

5. RESEARCH METHODOLOGY

The study adopts an Inductive Approach which is applied research from application view-point. It adopts a systematic approach using research methods and techniques, Quantitative and Inferential in nature, and the Research method is Descriptive and Applied. The methodology used is Action Research. The approach is Quantitative, and the method is Descriptive and Applied. Data is collected from Secondary sources, historical and current Secondary data for statistical analysis.

Research questions include how to achieve sustainable economic growth at constant GDP level, and whether there is an equilibrium point at which resource utilization from ecosystem equates with sustainable economic growth.

Research Method used is Statistical Data Analysis Techniques.

The study rests on the conceptual premise that Ecosystem utilization equals Economic Growth. The assumptions that follow are:

1. Universal GDP (UGDP) remains constant,
2. Economic Growth is sustained by Resource Utilization,
3. Resource depletion can be optimally managed by RGDP (Real GDP) modeling.

Purpose of Research: To design a robust sustainability model n alternate indicator to GDP for economic valuation and growth for significant inclusion of natural resources. It attempts to create a robust transparent mechanism for reporting utilization of natural resources with accuracy and transparency. The decision model based on constant cumulative GDP will ensure flow of excesses and shortfalls into the circular flow of world economy. All nations will get a fair chance at "equitable co-existence". The Prime Mission of the research is Economic Equality and Environmental Sustainability.

5.1 Research Problem and Research Questions

Economic Disequilibrium and Ecological imbalance forms the core of the problem which initiates this research. Natural resources are being indiscriminately exploited for mass production even when there is no apparent demand. A flawed system of accounting of natural resources is causing environmental degradation and threatens sustainability of the planet for future generations.

The fundamental questions to which the research study seeks answers are:

1. How can natural resources be included in economic growth indicator mechanism for environmental sustainability?
2. How to achieve equitable economic growth at constant GDP level?
3. Is there an equilibrium point between resource utilization and economic growth?

5.2 Research Aims and Objectives

Sustainability endeavors a call for managing equilibrium between optimal growth and scarce resources. It is the equilibrium point at which planned resource utilization sustains world economic growth. The objective is to formulate an econometric decision model to maintain constant equilibrium point between Economic Growth and Ecosystem sustainability. The aim of this research is primarily to discover the equilibrium point at which planned resource utilization sustains world economic growth. The main objective of the research is to formulate an economic growth based on valuation of natural resources, and to design econometric decision model to maintain equitable economic growth. The study deals with three Issues: Economic Equality, Environmental Sustainability, and Universal Ombudsman.

Laws of Pure Sciences Applied in the Context of this Study: The Law of Energy Conservation states that "Energy is neither created nor destroyed, but only transformed from one form to another". Since GDP is the sum total of the values of all the (energy) resources, it is postulated that total GDP of the Universe (UGDP) remains constant.

Energy can neither be created nor destroyed

As GDP is the sum total of the values of all the (energy) resources, and as resources can only be transformed from one state to another, it follows that,

The GDP of the Universe Remains Constant

This is the Law of Conservation of GDP.

Relationship between Variables: The relationship between the different variables is depicted in the figures below, which is self-explanatory:

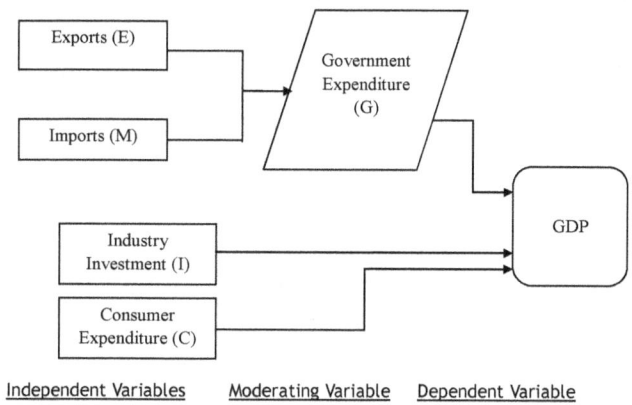

Fig. 5: Economy Related Relationship between Variables

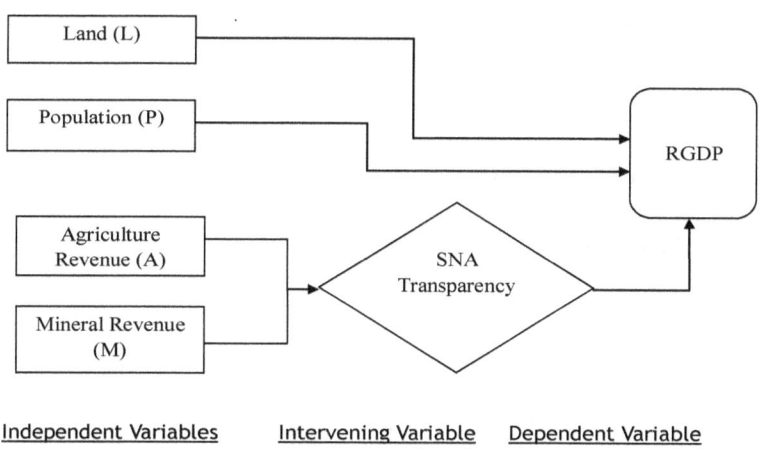

Fig. 6: Ecology Related Relationship between Variables

5.3 Statistical Data Analysis Techniques

> The basic idea proposed is: $GDP_{CUMULATIVE}$ EQUALS $RGDP_{CUMULATIVE}$

Data Analysis Techniques

1. For the purpose of deriving the Macro Constancy formula for Consistent Economic growth, Linear Models and Regression Modeling based on Factor Analysis and multivariate regression is used.
2. To derive Real GDP as an alternative indicator of national economic status, basic statistical techniques using means and averages, Standardization of Weights, and Weighted Index Method are used.

5.4 Originality of Research

This is an original comprehensive research that adopts a holistic approach to the Economy-Ecosystem conundrum, and offers a long-term ideal solution to the problem of sustainability.

6. ACTION PLAN

Business Model Innovation does not happen in vacuum. It happens only in the confines of the universe that mankind lives in. There is a need for an economic system that operates within our planetary boundaries. It is time to think out of the box and create the ultimate sustainable "Economic-Ecosystem Model" with validity lasting the life-time of the planet. This requires shifting foundations of current business models and incremental innovation for Economies to adapt and succeed, and to respect environmental limits while fulfilling social wants and needs. Innovation is the new mantra, but the innovation-invention gap, a transition from Invention to Innovation, called the "Valley of Death" or the "Darwinian Sea", will need to be negotiated. A radical paradigm shift is called for which entails encompassing the entire business model of the world to align to the needs of Equitable Economic Development with Environmental Sustainability as the underlying foundation.

Literature surveyed suggests that neither can big corporations restructure their established business models to accommodate rigorous sustainability demands, nor will the complete re-modeling of small businesses make any difference to the economic world. Hence it is necessary to rethink this paradox on the lines of Embeddedness.

Actors cannot change the environment that is shaping them unless there is collective participation from all agencies involved who will voice the same opinion. This entails entering the realm of collective entrepreneurship and rethinking the business models, not from any particular organizational point of view, but by putting all the business organizations big and small under one collective umbrella: the human world.

The Business entities here would mean the nations of the world and the spirit of business enterprise that we desire to inculcate and change through a sustainable business model would encompass the entire humanity into one collective business enterprise with a common goal. It is an exciting endeavor to encompass humanity in a spirit of entrepreneurial brotherhood and to bind humanity in an exciting business venture which will leave no scope for poverty, inequality or corruption; where environmental sustainability is assured for future generations inhabiting this planet, without compromising on the standards of living of the present.

Each nation is a profit center in this global venture. The World increasingly growing richer economically, but always limited by the targets set by the Valuation of the Natural resources deployed for the growth, and where target-achievement is followed by equitable distribution of the wealth among all the profit-centers. Citizens are also the internal consumers of the products of economic entrepreneurial enterprise. This spin-out to form a new entrepreneurial venture will enable citizen to escape the pathology of ideas by avoiding the familiarity, maturity, and nearness traps.

7. THE SUSTAINABLE ECONOMIC-ENVIRONMENTAL BUSINESS MODEL

The basic management tenets that design of the proposed "Sustainable Economic-Environmental Business Model" Incorporates are:

1. Organizational Restructuring
2. Business Process Re-Engineering
3. Strategic Business Unit Concept
4. Scientific-Economic Reprocessing
5. Holistic Logistical Integration

7.1 Highlights of the Model

1. The Business Model works on the underlying proposition of encompassing the entire globe into a spirit of entrepreneurship for the benefit of mankind and the planet.
2. For this purpose, the world consisting of all the nations is considered as one huge Business Enterprise with the nations as individual Strategic Business Units.
3. Each SBU will be responsible for ploughing in raw material (natural resources) into the processes of world economy–mostly manufacturing and agriculture.

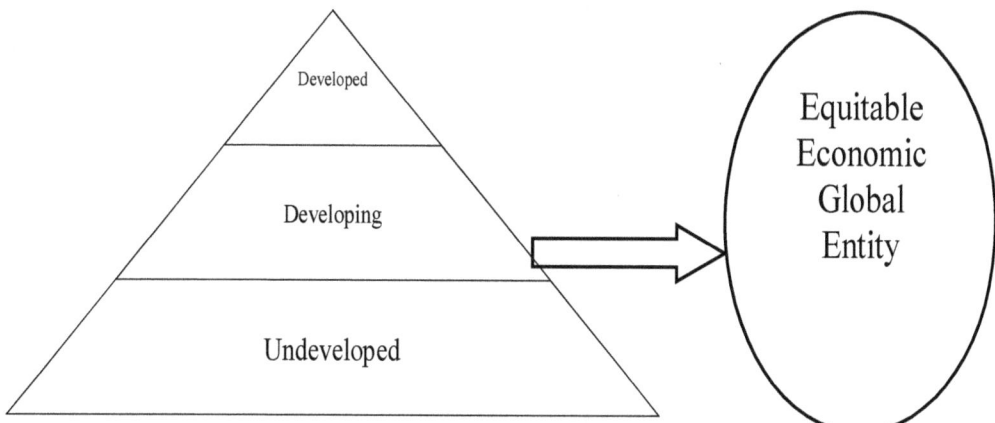

Fig. 2: Outline of a Global Business Enterprise

4. The ancillary services like transportation, infrastructure development, technology transfer etc would be shared responsibility between the SBUs.
5. Trade, commerce, imports, exports and similar activities between and within nations at macro-level will be regulated through the inter-mediation of a global ombudsman.
6. The key function of the Ombudsman would be regulation and monitoring of the natural resources utilized for economic purposes.
7. The final purpose is the equitable distributions of the wealth created by the commercial activities among all SBUs.

7.2 Traditional Vertically Integrated Utility Model *v/s* Wholesome/ Holistic Integration

Vertical integration is the degree to which a firm owns supply chain–its upstream suppliers and its downstream buyers, by bringing large portions of the supply chain not only under a common ownership, but also into one corporation. The characteristics of a traditional vertically integrated industry are Monopoly Franchise, Obligation to serve, and Regulatory Oversight.

*The nations encompass all three roles: C*ustomer in forward integration, supplier of resources in backward integration and ultimate consumers of the economic produce of economic activity: mixture of wholesome integration.

The resulting outcome unifying this triad is the World Business Organization which runs this Business Enterprise under the stewardship of the Ombudsman nation which will regulate, monitor and control not just the flow of natural resources, but also maintain equitable distribution of wealth.

Innovation for a change in ownership structure calls for not just changes in products and processes, but the entire Business/Economic model of the world. The Governments, Businesses and Society of all the nations will have to realign themselves to this new business model.

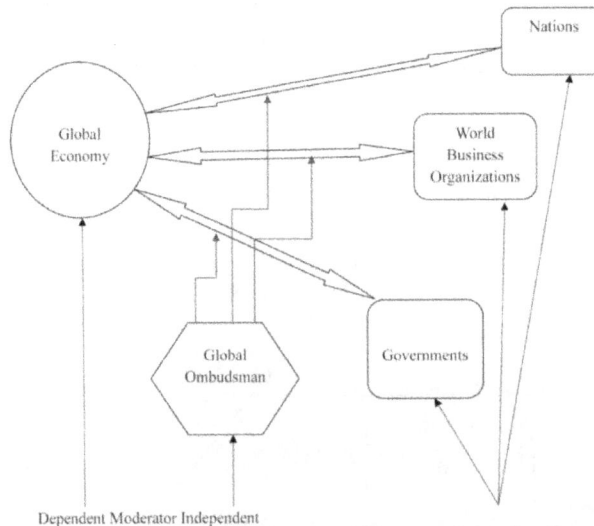

Fig. 3: Outline of Proposed Sustainability Model-Logistic Integration

Backward Integration will engulf the supplier end, which are the nations supplying natural resources for economic activities. Forward Integration consists of integrating the distribution channels for finished goods, which are the domestic and export consumption markets of the world. A step further in the holistic approach is the inclusion of the end-user, which are the citizen of every nation at micro-economic level. The citizen of the world are also equal owners of the natural resources of the world, which pushes the backward integration process one step further on the basis of proprietary ownership.

This is the Wholesome/Holistic Integration concept of the Economic-Ecosystem Equilibrium Model.

Introducing the Hypothesis: Universal GDP is a constant
Mathematically, for any nation,

$$GDP = C + G + I + NX$$

where C: all private consumption, G: sum of government spending, I: sum of all Businesses Outlays on capital, and NX : Net exports (Exports-Imports).

Extending this formula for the entire universe as a single unit comprising of all the nations, we have

$$UGDP(Y) = \beta + \beta_1 C + \beta_2 G + \beta_3 (X-M) + \beta_4 I + \mu$$

where UGDP (Y) = Universal GDP

Further regression proves that the single most important variable affecting UGDP with positive impact is Business Outlays (I). The equation regresses to:

$$Y = \beta + \beta_1 I + \mu$$

In the five sectors circular flow of income model, the state of equilibrium occurs when the total leakages are equal to the total injections that occur in the economy. This can be shown as:

Savings + Taxes + Imports = Investment + Government Spending + Exports

$$i.e. \ S + T + M = I + G + X$$

A similar line of assumption creates the equation for economic–environmental equilibrium. When Economic development equals the environmental resources utilized for its creation, the ideal equilibrium point is reached. This means we are utilizing exactly that amount of resources which are needed for economic growth, and also that economic growth is limited to only the amount of resources being utilized. The time and spatial dimensions are relevant issues.

The fundamental assumptions in this research are:
(a) Natural Resources of the world are commonly owned by all nations
(b) Real Gross Domestic Product (RGDP) is the true indicator of economic growth
(c) When RGDP equals constant GDP, the balance between Economic and Environmental. Growth is reached. This is the ideal Equilibrium point.

Conceptualizing the Six Sector Model of Economy

The Core Idea behind the Six-Sector Model of Economy is the introduction of the Global Ombudsman at macro level as a facilitator to the World Economy.

The various types of variables in a six sector model are:

- *Dependent variable:* GDP
- *Independent variables:* Investments, Consumption, Export, Import
- *Moderating variables:* Govt. Expenditure
- *Mediating variables:* Global Ombudsman.

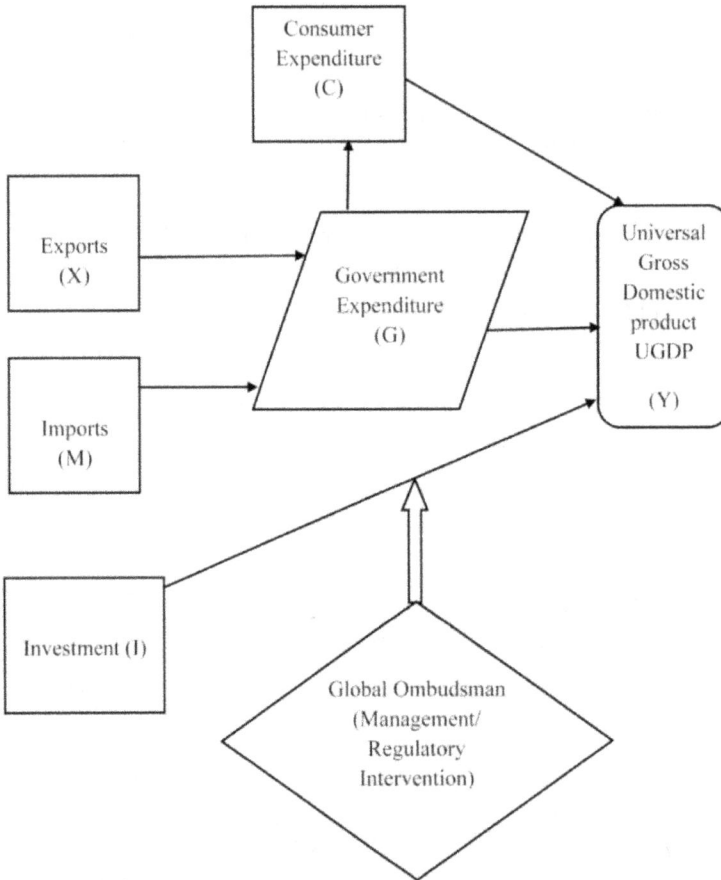

Fig. 4: Role of Ombudsman–Regulatory and Moderating

Inference Derived from the Ombudsman Model

There is scope for intervention by the Global Ombudsman body as a Mediating Variable, by exercising interventionist influence on the utilization of natural resources and regulatory role of re-distribution of economic wealth equally among all nations of the world.

Innovative Financing Model

The finances of the world created out of economic activity, global trade and inter-country transactions are to be ploughed back into the circular flow of world economy after equitable distribution of the surplus wealth. Surplus wealth is defines as the excess economic value created over the real value of environmental resources utilized for economic purpose.

At macroeconomic level, the current linear "Take, Make, Dispose" economic system of extracting resources from the planet and producing use-and-dispose products is a very unsustainable approach. A circular economy is waste-free and restorative of ecosystems, closing resource loops, always taking into account the social and ecological impact.

An efficient Economy is one in which materials streams are efficiently managed and recycled, runs entirely on the basis of renewable energy without negative effects on human life or the ecosystem. There is reformatory role of money and finances from traditional production and consumption of goods and services, by a total revamp of economic performance measurement tools. It optimizes systems rather than components, and considers substituting manpower for energy. The traditional concept of the circular economy is the circulation of money verses goods, services, etc.in macroeconomics. In the new sustainability concept, the flow is in only one direction, till the complete redistribution of recycled goods over the world is achieved.

Flow of Sustainable Business Modeling Process

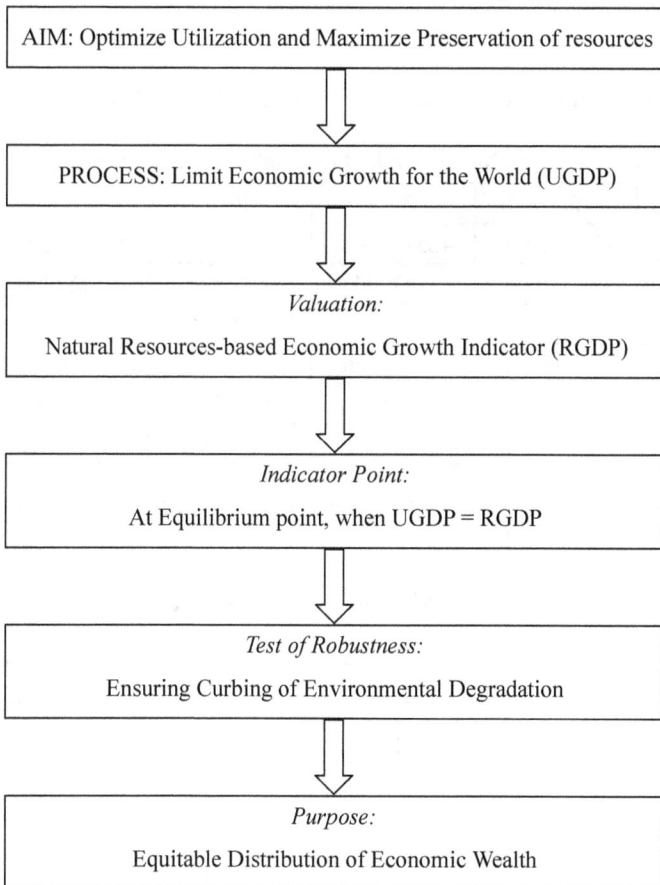

AIM: Optimize Utilization and Maximize Preservation of resources

⇩

PROCESS: Limit Economic Growth for the World (UGDP)

⇩

Valuation:

Natural Resources-based Economic Growth Indicator (RGDP)

⇩

Indicator Point:

At Equilibrium point, when UGDP = RGDP

⇩

Test of Robustness:

Ensuring Curbing of Environmental Degradation

⇩

Purpose:

Equitable Distribution of Economic Wealth

8. MAIN BODY OF RESEARCH

The Sustainability Business Model is derived through a three-stage process:
1. Deriving Real GDP (RGDP) for Economic Valuation of Natural Resources
2. Calculating cumulative GDP (UGDP) of nations
3. Arriving at the Economic-Ecosystem Equilibrium.

Stage 1: Deriving Real GDP Index (in RGDP Units)

Data Analysis and Statistical Technique Used:

Using basic statistical techniques using means and averages, Standardization of Weights, and Weighted Index Method, Real GDP is derived. The four basic parameters that define a nation's economy are Land, Population, Agriculture (revenue), and Minerals (revenue).

Statistical inference for correlation between GDP (dependent variable) and Land, Agriculture, Population, and Minerals (independent variables) suggests that GDP is dependent most on revenue from minerals.

Derivation Process

Step 1: The mineral output for the 12 nations for period 2009 to 2013 is collected.

Table 7: Mineral Output for Australia (2009–2013)

Commodity	2009	2010	2011	2012	2013
Coal (mT)	418252000	425751000	416733000	430124000	477000000
Silver (kg)	1702000	1879000	1725000	1728000	1840000
Gold (kg)	224000	260000	259000	253000	265000
Steel (mT)	5235000	7408000	46538000	4904000	4645000
Petroleum (MT)	22527000	24796000	21967000	21814000	18194000
Uranium (T)	7942	5971	5961	7022	6432

Step 2: The Relative Weight is calculated keeping 2009 as base Year, and the price of each commodity is taken as standard.

Table 8: Relative Weight of Commodities (Minerals)

Relative Wt	2009	2010	2011	2012	2013
Coal	100	102	98	103	111
Silver	100	110	92	100	106
Gold	100	116	100	98	105
Steel	100	142	628	11	95
Petroleum	100	110	89	99	83
Uranium	100	75	100	118	92
Arithmetic mean	100	109	184	88	99

Table 9: Standardized Price of the Minerals

Current Int'l Prices			Standardised			Standardised price/kg	
Coal	$51.68	st	6.3503	Kg		$8.14	Kg
Silver	$15.57	per ounce	0.02835	Kg		$549.22	Kg
Gold	$1,195.95	per ounce	0.02835	Kg		$42,185.94	Kg
Steel	$100.00	T	1000	Kg		$0.10	Kg
Petroleum	$64.00	barrel	0.45	Kg	125 Lb	$140.80	Kg
Uranium	$37.00	per lb	0.45	Kg		$82.22	Kg

Important Note: It is proposed that since all minerals belong commonly to mankind, they should have a standard price at any place on earth at any point in time.

Step 3: The Weighted Index and the Standardized Weighted Index for the 6 Minerals are calculated.

Table 10: Weighted Index of the Six Minerals

Weighted Index	2009	2010	2011	2012	2013
	100	44	95	100	105

Step 4: The Standardized Mineral Revenue and Average Annual Revenue are derived.

Table 11: Standardization of Revenue Weighted Index

Standardization of Revenue for Australia (USD)						
Std. Weight	2009	2010	2011	2012	2013	Average
Coal	21615263	22002812	21536761	22228808	24651360	22407001
Silver	26500140	29256030	26858250	26904960	28648800	27633636
Gold	267892800	310947000	309751050	302575350	316926750	301618590
Steel	523500	740800	4653800	490400	464500	1374600
Petroleum	1441728	1586944	1405888	1396096	1164416	1399014
Uranium	293854	220927	220557	259814	237984	246627
				Avg. Annual Revenue		$ 354679469
						USD 355 mn

Step 5: Final derivation of Real GDP

1. Using the data collected for land, population, agriculture revenue and mineral revenue (as derived above), the economic growth of the nations is thus calculated to arrive as RGDP (Real GDP) as the new indicator of economic growth.

2. Population is considered as a multiplicative component as it assists circular flow of economy exponentially through consumerism.
3. The Land Output is considered as combination of both agriculture and mineral revenue.

Step 6: Final Output of Real GDP for 12 Nations–Analytical Purpose

The final output for the 12 nations for time period of five years in terms of Real GDP (RGDP) is tabulated as follows:

Table 12: Summary of Economic Status of 12 Nations (RGDP-Based)

Parameters	Symbol	{Units}	Australia	Brazil	Canada	China	Germany	India
Population	X	mn	22.74	198.64	34.76	1350.84	81.11	1236.60
Land	L	sqkms	7682300	8418908	9093510	9351778	348566	2973190
Agri Revenue	Z1	mn USD	26257	111324	21092	705432	24095	210680
Min Revenue	Z2	mn USD	354.68	82.27	151.09	750.40	26.14	46.04
Land Output	O = Z1 + Z2	mn USD	26612	111407	21243	706182	24122	210726
Rev/Land	Y = O/L	mn $/Sq Km	0.0035	0.0132	0.0023	0.0755	0.0692	0.0709
Real GDP	**RGDP = X*Y**	**RGDP**	**0.08**	**2.63**	**0.08**	**102.01**	**5.61**	**87.64**

Parameters	Symbol	{Units}	N. Z'land	Russia	S. Africa	Spain	UK	USA
Population	X	mn	4.42	143.27	52.37	46.62	63.67	314.11
Land	L	sqkms	263310	16376870	1213090	498800	241930	9147420
Agri Revenue	Z1	mn USD	9665	40084	5912	30039	16013	197413
Min Revenue	Z2	mn USD	14.92	311.86	225.68	3.16	5.50	375.86
Land Output	O = Z1 + Z2	mn USD	9680	40395	6137	30042	16018	197788
Rev/Land	Y = O/L	mn $/Sq Km	0.0368	0.0025	0.0051	0.0602	0.0662	0.0216
Real GDP	**RGDP = X*Y**	**RGDP**	**0.16**	**0.35**	**0.26**	**2.81**	**4.22**	**6.79**

Implication of Derivation of RGDP: The Real GDP values for these 12 nations give an indication of their Economic Strength based on Valuation of Natural resources.

Stage 2: GDP as a Sum Aggregate of Economic Activities

Gross Domestic Product (GDP) is the monetary value of all goods and services, including all private and public consumption, government outlays, investments and foreign trade in a country. Mathematically, GDP = C + G + I + NX…. where C: Consumer Expenditure, G: Govt. spending, I: Industry Investment, NX: Net of Imports and Exports

As a corollary, Universal GDP is defined as the sum aggregate of the monetary values of the resources of all the nations of the world. *i.e.*

UGDP = Market Value of Resources + Income of People + Govt. Spending + Foreign Trade

Table 13: List of Sample Nations for Cumulative UGDP

Country Name	
Australia	New Zealand
Brazil	Russia
Canada	South Africa
China	Spain
Germany	UK
India	USA

The data based on economy-related parameters is collated as follows:

Table 14: Economic Data for 12 Nations

Economic Data : 2009–2013 Average (USD Mn)				
Country Name	GDP	Govt. Exp	Net Trade	Cons. Exp
Australia	1079826	399104	−4637	590679
Brazil	2119388	806375	−27865	133856
Canada	1378513	579025	−27127	751367
China	7372272	538353	218456	7372272
Germany	2977265	1353735	189979	1628329
India	1166141	318474	−88846	605122
New Zealand	141382	60642	2462	79335
Russia	972204	379732	129316	493297
South Africa	232623	30276	−703	140062
Spain	1196033	546499	5739	677009
UK	2488275	1181188	−49359	1528017
USA	15566450	6392934	−512940	10377099

Note:

1. Wealth Creation is the primary motive for business, but *Unequal Wealth Distribution* causes unbalanced economic growth.
2. The concept of Universal Gross Domestic Product (UGDP) is a measure to counter business cycles for a stable world economy. Natural Resources are the real wealth of nations.

3. As nations under-report their role in economic growth, *optimization strategy* for intelligent utilization of natural resources is suggested, using regression analysis techniques.

Simultaneous Equation Model: As there is more than one dependent variable, we have $Y = \alpha + \beta [C + G + (X–M) + I] + \mu$

Multiple Regression Analysis: Each function will be related with one dependent variable into regression, which includes non-linearity. Thus the equation regresses to $Y = \beta + \beta 1\ C + \beta 2\ G + \beta 3\ (X – M) + \beta 4\ I + \mu$, where C: Consumer Expenditure, G: Government Expenditure; X: Exports, M: Imports:

$C = \alpha + \beta$ [per capita Income] $+ \mu$

$G = \alpha + \beta$ [tax rate and excise] $+ \mu$

$X = \alpha + \beta$ [production output] $+ \mu$

$M = \alpha + \beta$ [domestic demand] $+ \mu$

$I = \alpha + \beta$ [Capital Formation/FDI] $+ \mu$

To summarize, the Average Co-relation Variance calculated for an earlier study conducted by the same author is as follows:

Table 15: Correlation Statistics

	GDP v/s C	*GDP v/s G*	*GDP v/s I*
Average correlation	0.9852358	0.993358	0.8591405

Conclusion: There is more scope to influence GDP by focusing attention on Capital Industry Investments.

8.1 Universal GDP is a Constant

Mathematically, for any nation, GDP = C + G + I + NX, where C: all private consumption, G: sum of government spending, I: sum of all Businesses Outlays on capital, and NX: Net exports (Exports-Imports). Extending this formula for the entire universe as a single unit comprising of all the nations, we have

$UGDP(Y) = \beta + \beta 1C + \beta 2G + \beta 3(X–M) + \beta 4I + \mu$, where UGDP (Y) = Universal GDP.

Considering the universe as one single business entity of all nations, we have X = M, and X–M = 0. Thus, $Y = \beta + \beta 1C + \beta 2G + \beta 3I + \mu$. Also, ignoring the influence of Consumer and Government expenditure, the equation regresses to $Y = \beta + \beta 1I + \mu$. We may thus conclude that Capital Investment in industries (I) has maximum influence on UGDP.

8.2 Scope for Business Decision

Based on the Analysis of Descriptive Statistics, conclusions for two sample nations Australia and India are produced below:

Table 16: Sustainability Model - Pilot Study for India and Australia

Australia	*2009*	*2010*	*2011*	*2012*	*2013*
GDP	958838	1041820	1101141	1125499	1171832
Average GDP:	**1079826**				
Consumer Expenditure (C)	530962	561925	591550	618949	650009
Net Trade (N)	−9895	6640	13079	−22942	−10065
Govt. Spending (G)	366660	383181	407202	409119	429359
a = C + N + G	887727	951746	1011831	1005125	1069304
Investment (I) = (GDP – a)	71111	90074	89310	120374	102529
India	*2009*	*2010*	*2011*	*2012*	*2013*
GDP	848571	977424	1159818	1342671	1502220
Average GDP:	**1166141**				
Consumer Expenditure (C)	605122	605122	605122	605122	605122
Net Trade (N)	−60107	−84095	−102633	−118204	−79193
Govt. Spending (G)	239891	266055	315470	365206	405750
a = C + N + G	784906	787083	817960	852125	931678
Investment (I) = (GDP – a)	63665	190342	341858	490546	570542

For any decision based on the decision framework model, the government or any appointed ombudsman may decide to inject or withdraw from the common pool.

Table 17: Business Decision Frame Work Based on Sustaianability Model–Pilot Study for India and Australia

India	*2009*	*2010*	*2011*	*2012*	*2013*
Shortfall (Avg. GDP – Actual GDP)	317570	188716	6323	−176530	−336079
Action for Shortfall/Excess	Injection	Injection	Injection	Withdrawal	Withdrawal
Australia	*2009*	*2010*	*2011*	*2012*	*2013*
Shortfall (Avg. GDP – Actual GDP)	120988	38006	−21315	−45673	−92006
Action for Shortfall/Excess	Injection	Injection	Withdrawal	Withdrawal	Withdrawal

Note:

1. The tables show the Average GDP for 2 sampling units out of sample size of 12 nations, along with the shortfall or excess in Investment Expenditure.
2. The average GDP figure can be replaced by any other convenient figure, for e.g. a percentage of the RGDP (total GDP based on valuation of natural resources using the method described in Part A).

Stage 3: State of Economic-Ecosystem Equilibrium

In an economy, a stable state of equilibrium is reached when the total leakages are equal to the total injections: $S + T + M = I + G + X$

Savings + Taxes + Imports = Investment + Government Spending + Exports

In the stable state, when wealth from economic activity equals natural resources utilization, the equilibrium point between economic growth and ecosystem sustainability is reached. Mathematically, this equilibrium point is represented as:

$L + P + A + M = C + G + I + N$

(i) For time period 2009 to 2013, for the 12 nations, GDP $_{cum} = C + G + I + N$

Table 18: Cumulative GDP for 12 Nations

GDP $_{cumulative}$		C	G	I	N
36690370	=	12586338	−165524	24376444	−106887

(ii) The Real GDP for the same time period is calculated as follows:

Table 19: Real GDP for 12 Nations

Real GDP		Agriculture Rev		Mineral Rev		Population
1310649234	=	1398005	+	2348	*	3549

(iii) The detailed working for RGDP is given in tabular form below:

Table 20: Working for RGDP Derivation

Country	RGDP Working						RGDP	
	Revenue (USD mn)			Land	Rev/Sq Km	Popln Mn	Ttl Rev * Pop	Rev/Km * Pop
	Agricultural	Mineral	Total	Sq Km				
Australia	26257	355	26612	7682300	0.0035	23	605239	0.079
Brazil	111324	82	111407	8418908	0.0132	199	22129771	2.629
Canada	21092	151	21243	9093510	0.0023	35	738400	0.081
China	705432	750	706182	9351778	0.0755	1351	953936241	102.006
Germany	24095	26	24122	348566	0.0692	81	1956428	5.613
India	210680	46	210726	2973190	0.0709	1237	260584345	87.645
N. Z'land	9665	15	9680	263310	0.0368	4	42774	0.162
Russia	40084	312	40395	16376870	0.0025	143	5787360	0.353
S. Africa	5912	226	6137	1213090	0.0051	52	321401	0.265
Spain	30039	3	30042	498800	0.0602	47	1400683	2.808
UK	16013	6	16018	241930	0.0662	64	1019861	4.216
USA	197413	376	197788	9147420	0.0216	314	62126730	6.792
Total	**1398005**	**2348**	**1400353**	**65609672**	**0.0213**	**3549**	**1310649234**	**75.752**

Inferences and Recommendations Based on Results of Decision Model

When we equate the RGDP value (Table 20) with cumulative GDP (Table 18), we see that RGDP is about 36 times more than the $GDP_{cumulative}$.

1. This gives us a rough estimate of the amount to be replaced in the Decision Model for Economic Growth activity.
2. The total amount to be capped for natural resource utilization for a thirty year period is equivalent to the RGDP value.
3. The total annual value for economic activity of a nation to be targeted for a future growth period can be determined keeping the RGDP value constant.
4. This will ensure sustainable economic growth, along with environmental stability.

9. SIGNIFICANCE OF THE STUDY

This study offers a solution to the problem of sustainability. The recommendations that are derived from the models are simple, workable solutions that require a developmental mindset. Economic progress based on inclusion of natural resources in growth mechanism will contribute to a better world. Redistribution of wealth into the circular flow of economy will bridge the "Economic Divide", and offer better growth opportunities to poor nations.

10. SCOPE AND LIMITATIONS

Scope: When natural resources are commonly pooled under universal proprietorship, national boundaries will be obliterated. Real time evaluation of natural resources will enhance growth in the science of the measurement and valuation by timely, accurate and reliable data collection, assimilation, and their economic analysis for each nation that contributes to world economy.

Limitations: The theoretical framework limits the study to

1. Focus on specific variables,
2. Sampling size,
3. Specific viewpoint
4. Data Procurement Limitations,
5. Time and resource constraints, and
6. Acceptance of the Model in practical economic occurrence by governments of nations.

11. MANAGERIAL IMPLICATIONS

1. The study offers a decision model for all nations based on sustainability principles through a constant cumulative GDP for equitable growth. All nations can hope for equal opportunity is economic growth.
2. An alternative to GDP as a growth indicator is offered by valuation of natural resources in the form of Real GDP (RGDP). This can assist managerial decision-making process to form policies and evaluate progress, in order to promote truly sustainable development while supporting the ecosystem.

3. Real time evaluation of natural resources will enhance the scope for business in the science of the measurement and valuation.

4. Efforts of environmentalists will be directly translated to wealth creation through ecology sustenance and sustainable growth.

5. The role of India as a leader in regulating utilization of natural resources of the world will enhance our prospects towards a strong and resilient world economy.

11.1 Role of Government

Government's role in policy formulation involves facilitating distribution of global economic wealth based on the decision model under the control of a universal ombudsman. Governments can also execute an efficient system of national accounting of natural resources within demographic boundaries with a sense of only custodianship, and participatory universal proprietorship.

11.2 Assumptions

1. Natural Resources are common property of mankind, hence the revenue derived from their utilization must be shared equally among all nations

2. Natural Resources are limited; hence their utilization must be limited depending upon the need for economic growth.

3. Nations are custodians and not propriety owners of the natural resources. To ensure their role in regulated utilization of natural resources and equal distribution of economic spoils, a universal ombudsman to oversee the proceedings.

11.3 Managerial Actions Suggested for SBU Nations

1. Any nation needing resources, can freely take up from the common pool.

2. Any Nation exceeding GDP growth above the threshold limit for nations must pool back the excess amount into the common pool.

3. To ensure smooth running of the mechanism, a nations acting as a global ombudsman needs to be appointed who can regulate and control the utilization of natural resources and distribution of economic wealth.

12. RESEARCH FINDINGS

Expected Outcomes of Research include Economic Equilibrium Model, and Ecosystem Sustainability Model, with Business Opportunity for Environmental Consultancy, and Growth Opportunity for Science of Calibrations and Measurements.

The Mission of the Sustainable Economic-Environmental Business model is put a check on environmental degradation by preventing rapid, unplanned depletion of natural resources, and ensure equitable development of all nations of the world irrespective of their contribution to world economy by ensuring that the wealth created out of commercial activities of nations are pooled in a common treasury, and distributed equally among all nations.

This will prevent the race for power among nations based on economic superiority, and resolve the problems of the world amicably.

13. CONCLUSIONS

This Economic–Environmental Equilibrium Sustainability Business Model is the decision model that would change the face of business the world over, and turn the economics of the world upside down. There will never be a more sturdy and robust Model of Business Sustainability. To ensure this happening, India is the only hope for mankind for sustainable development of the world economy, in the role of an able ombudsman.

For India, taking up the challenge of a global leadership is just the first step towards Economic Supremacy–the dream of the missile-man.

BIBLIOGRAPHY

Bluman, A.G. (2009). Elementary Statistics. McGraw-Hill, New York, NY.

Dwiwedi, D.N. (2005). Macroeconomics–Theory and Policy. Tata McGraw-Hill Education.

Ekaran, U. (2002). Research Methods for Business. Wiley.

Hair, J. *et al.* (2009). Multivariate Analysis. England: Pearson Higher Education

Keynes, J.M. (1965). General Theory of Employment, Interest, and Money. Harcourt Brace, New York.

Koutsoyiannis, A. (1972). Theory of Econometrics. 2nd edition, Macmillan Education Ltd., Econometric Methods, Johnston J., 1997, McGraw-Hill.

Levin, R.I. and Reuben, D.S. (2008). Statistics for Management. 7th edition, Prentice Hall of India Pvt.

Ramanathan, R. (2002). Introductory Econometrics with Applications. Harcourt College Publishers.

Saunders, M., Lewis, P. and Thornhill, L. (2009). Research Methods for Business Students. Fifth, Pearson, England.

Ulbrich, M. (1989). Introduction to Economic Principles. McGraw-Hill.

Verma, L. and Mittal, G.S. (2012). Introduction to Macro Economics. S. Dinesh and Co.

REFERENCES

[1] Hannesson, R. (2009). "Energy and GDP growth", *International Journal of Energy Sector Management,* Vol. 3, Issue 2, pp. 157–170.

[2] Al-Mulali, U. (2014). "GDP growth—Energy Consumption Relationship: Revisited", *International Journal of Energy Sector Management*, Vol. 8, Issue 3.

[3] Repetto, R. "Earth in the Balance Sheet: Incorporating Natural Resources in National Income Accounts", *Environment* (Sep 1992), Volume 34, Issue 7, pp. 1–12.

[4] Aguado, R. and Martinez, J. (2012). "GDP and Beyond: Towards New Measures of Sustainability based on Catholic Social Thought", *Asia-Pacific Journal of Business Administration,* Vol. 4, Issue 2, pp. 124.

[5] Lyakurwa, W. (2009). "Prospects for Economic Governance: Resilient Pro-poor Growth", Foresight, Vol. 11-4, pp. 66.

[6] Urhammer, E. (2014). "Crisis in the Habitat of the Economic Growth Monster", On the Horizon, Vol. 22.

[7] Rajasingham, C. (1998). "Alternative Indicators for Development: A Case Study of Atlantic Canada", Thesis, Saint Mary's University, Halifax, Nova Scotia, Canada.

[8] Al-Rawashdeh, *et al.* (2014). "Air Pollution and Economic Growth in MENA Countries", Environmental Research, Engineering and Management, 4(70): 54–65.

[9] Perrings, C. (1998). "Resilience in the Dynamics of Economy-Environment Systems", *Environmental and Resource Economics*, April 1998, Volume 11, Issue 3–4, pp. 503–520.

[10] Hobijn, B. and Steindel, C. (2009). "Do Alternative Measures of GDP Affect Its Interpretation?", *Current Issues in Economic and Finance,* Vol. 15, No. 7, Nov. 2009.

[11] Boedo, M. and Hernan, J. (2006). "Three Essays in Dynamic Macroeconomics", *ProQuest*, 2006.

[12] Zsolnai, L. (1993). "A Framework of Alternative Economics", *International Journal of Social Economics*, Vol. 20, Issue 2, pp. 65–75.

[13] Dave, B. (2003). "Money v/s Wealth: Why GDP is the Wrong Measure?", *Mediaweb Ltd., Profile Publishing Ltd, Auckland*, New Zealand.

[14] Tschirhart, J. (2009). "Integrated Ecological- Economic Models", *Annual Review of Resource Economics*.

Financial Feasibility of CDM Projects

Renuka H. Deshmukh[1], Snehal Nifadkar[2] and Anil P. Dongre

[1]Department of Management, International Institute of Management Studies, Pune
[2]Minda Stoneridge Ltd., Pune
[3]Department of Management, North Maharashtra University, Jalgaon
E-mail: [1]renuka.nifadkar@gmail.com; [2]snifadkar@gmail.com;
[3]ap_dongre@rediffmail.com

ABSTRACT: *The research study aims to analyze the financial performance of the companies associated with CDM projects implemented in India from 2001 to 2014 by calculating net profit with and without CDM revenue. Further the study also highlights the Year-wise and sector-wise lending to CDM projects in India as well as in the state of Maharashtra. The study further aims to examine the year-wise trend of Certified Emission Reductions (CER) issued by the CDM projects implemented in Maharashtra from 2001–2014. The study as well analyses the responses of selected corporate with respect to the challenges in implementing and obtaining finance from commercial banks.*

Keywords: Adaptation Costs, Internal Rate of Return, Mitigation, Vulnerability, CER.

1. INTRODUCTION

Corporate sector play a significant role in shifting investments to renewable energy production and energy efficiency solutions, and away from high-carbon and fossil fuel investments. Corporate sector can integrate climate risk into overall client risk identification and assessment process and develop a set of assessment tools to determine carbon reduction options. Challenges for the Corporate sector appears from diverse directions such as: regulations that are planned to limit GHG emissions, physical changes that take place due to climate change impacts, legal challenges to be brought on by insufficient governance, reputational fallout for companies due to corporate positions on climate change, and cut throat pressures in the marketplace as production costs shift and products are substituted in response to the new reality of a carbon-constrained world (Pearce, 2005). Indeed, reports have been published forewarning of the budding exposures in all segments of the sector (AGECCC, 2006) (Lloyd's, 2006).

These challenges present new risks and opportunities for financial sector and for the financial service industry as well (Kalia, 2011). The sector will have to acclimatize internal policies, processes, products, and services, in order to meet the challenges that its clients face and to defend its own feasibility. At the same time, climate change will unlock new opportunities for the financial sector, one of which is carbon financing under Clean Development Mechanism (CDM).

CDM under the Kyoto protocol intends at a cost-effective cutback of GHG emissions, technology know-how and capital movement from industrialized to emergent and developing countries (Nagai, 2005).

The present execution of the CDM has been concerned and extensively questioned about (i) the environmental efficiency that relates to whether the CDM contributes to global greenhouse gas emission reductions (Cames, 2003) (Michealowa, 2007) (Schneider, 2007) (Chung, 2007) (Victor, 2008) (Wara, 2007) (Schatz, 2008) (Haya, 2009) (Ghosh, 2007) (ii) restricted capacity to obtain financing for the fundamental greenhouse gas emission reduction activities, predominantly in the least developed countries, inadequate or no awareness of the CDM Modalities and Procedures among financial intermediaries in the CDM host countries, deficiency of enough approaches, tools and skills for CDM project evaluation or either are uneven and irregular to the skills in analogous institutions in developed countries, lack of experience in structuring arrangements for financing a project among potential project proponents (Kamel, 2007) (iii) its contribution to the sustainable growth goals in host nations as it's formerly framed in the Article 12 of the Kyoto Protocol (Olsen, 2008) (Matschoss, 2007) (Sutter C., 2007) (Brown, 2004) (Lohmann, 2006) (Sirohi, 2007) (UNEP, 2010) (iv) the need to encourage equitable geographic allocation of the CDM project activities at regional and sub regional levels (Cosbey, 2007) (Bakker, 2009) (v) the institutional arrangement of the CDM and connected hurdles in implementing CDM project activities like long project cycle, lead times and estimated CER creation, additionality argumentation, stakeholder involvement *etc.* (Michaelowa, 2003) (Michaelowa A.F., 2005) (Sterk, 2006) (Streck C., 2007) (Michaelowa A.A., 2008) (Bohringer, 2005) (vi) sectoral allotment and windfall profits of projects that relates to the discrepancy of GHG reduction by CDM projects among sectoral emissions reductions potential (Zegras, 2007) (Mendis, 2004) and high windfall profits/producer surplus of some project proponents and host countries as the costs of achieving some emission reduction have been very low (Grubb, 1999) (Kolshus, 2001); Kolshus, Hans H. *et al.* (2001) (vii) Lack of knowledge among projects entities, local banks and insurance companies (Thorne, 2008) (Curnow, 2009).

But it has experienced a strong accomplishment and steady growth since 2005:

As of May 2015, **all over the world** total 8630 CDM projects are now included in the Pipeline, out of which major share is occupied by renewable energy sector *i.e.* 6128, excluding the 267 projects given a negative validation by DOEs. 7630 of the projects are now registered and a further 14 are in the registration process. 2770 CDM projects have got CERs issued. (UNEP Risoe CDM/JI Pipeline Analysis and database)

In India, as of May 2015, total 2211 CDM projects are registered, out of which major share of 930 projects is occupied by wind energy followed by 231 Hydro Projects.

In the state of Maharashtra as of 31st December 2014, total 304 CDM projects are registered, out of which wind sector has prime share of 162 projects followed by 12 hydro projects. (http://cdmpipeline.org/publications/CDMStatesAndProvinces.xlsx)

On this background, this research study analyses the responses of selected corporates with respect to the challenges in implementing and obtaining finance from commercial banks.

2. METHODOLOGY

- *Research Design:* Descriptive.
- *Sample Unit:* Private/Public entities associated with CDM projects implemented in Maharashtra.
- *Data Sources and Sample Size*
  1. *Primary Data:* Primary data is collected through questionnaire and interview method from Private/Public entities associated with CDM projects.
  2. *Secondary Data:* Final Private/Public entities associated with CDM projects were selected from https://cdm.unfccc.int/index.html and from Project Design Document (PDDs) of individual CDM projects.

The financial data for analyzing the financial feasibility of CDM projects was obtained from https://cdm.unfccc.int/index.html and National CDM Authority Ministry of Environment and Forests, Government of India www.cdmindia.gov.in. 20 CDM projects in India were randomly selected.

Data for calculating Certified Emission Reductions (CER) by the CDM projects was obtained from https://cdm.unfccc.int/index.html -monitoring reports of individual CDM project. 10 CDM projects in Maharashtra were randomly selected for study.

Data regarding current and global scenario of CDM projects was collected from UNEP DTUCDM/JI Pipeline Analysis and Database www.cdmpipeline.org

Data regarding lending to CDM projects was collected from the year wise annual reports of IREDA and PFC.

3. ANALYSIS

3.1 Financial Analysis of 5 Selected CDM Projects in India

Financial Analysis of CDM Projects in India

Project 1: 12 MW Bundled Wind Power Project in Tenkasi, Tamilnadu (0796).

Brief Description
The project activity is the installation of 16 Wind Electric Generators (WEGs) in Tenkasi of Tirunelveli District in Tamil Nadu, Southern India.

Table 1: Financial Analysis without CDM Revenue (₹ in Lacs)

		2003	2004	2005	2006	2007	2008
1.	**Total Revenue (without CDM)**	927	927	927	927	927	927
2.	**Less: Operational Cost**						
	– Operation and Maintenance Cost		71	76.1	81.6	87.4	93.7
	– Insurance	18.3	18.3	18.3	18.3	18.3	18.3
	Total Expenses	18.3	89.3	94.4	99.9	105.3	112
3.	PBIDT (1–2)	909	838	833	827	821	815
4.	Depreciation	244	244	244	244	244	244
5.	Interest	244	244	195	147	98	49
6.	PBT [3–(4 + 5)]	420	349	393	436	479	522
7.	Tax	33	27	31	34	38	41
8.	**PAT** (6–7)	387	322	362	402	442	481
9.	Add Depreciation	244	244	244	244	244	244
10.	**Net Cash Accruals (Cash Inflow)** (8 + 9)	631	566	606	646	686	725
11.	**Less: Use of Funds (Cash Outflow)**						
	Capital Expenditure (Debt– 1921 + Equity– 2961 = **4882**)						
	Repayment of Term Loan		592	592	592	592	592
	Total Cash Outflow:		592	592	592	592	592
12.	Surplus/Deficit (10–11)	631	(26)	14	54	94	133

Continue...

		2009	2010	2011	2012	2013	2014
1.	**Total Revenue (without CDM)**	927	927	927	927	927	927
2.	**Less: Operational Cost**						
	– Operation and Maintenance Cost	100.5	107.7	115.4	123.7	132.6	142.1
	– Insurance	18.3	18.3	18.3	18.3	18.3	18.3
	Total Expenses	112	118.7	126	133.7	142	150.9
3.	PBIDT (1–2)	808	801	793	785	776	766
4.	Depreciation	244	244	244	244	244	244
5.	Interest						
6.	PBT [3–(4 + 5)]	564	557	549	541	532	522
7.	Tax	44	44	43	42	42	41
8.	**PAT** (6–7)	520	513	506	498	490	481
9.	Add Depreciation	244	244	244	244	244	244
10.	**Net Cash Accruals (Cash Inflow)** (8 + 9)	764	757	750	742	734	725
	Less: Use of Funds (Cash Outflow)						
	Capital Expenditure (Debt- 1921 + Equity- 2961 = **4882**)						
	Repayment of Term Loan						
	Total Cash Outflow:						
12.	Surplus/Deficit (10–11)	764	757	750	742	734	725
	IRR Without CDM Revenue	9.49%					

Source: Researcher's own study.

Table 2: Financial Analysis with CDM Revenue (₹ in Lacs)

		2003	2004	2005	2006	2007	2008
1.	**Total Revenue (without CDM in lacs)**	927	927	927	927	927	927
	CDM Revenue	158	158	158	158	158	158
	Total Revenue (with CDM revenue in lacs)	1085	1085	1085	1085	1085	1085
2.	**Less: Operational Cost**						
	– Operation and Maintenance Cost		71	76	82	87	94
	–Insurance	18	18	18	18	18	18
	Total Expenses	18	89	94	100	106	112
3.	PBIDT (1–2)	1066	995	990	985	979	973
4.	Depreciation	244	244	244	244	244	244
5.	Interest	244	244	195	147	98	49
6.	PBT [3–(4 + 5)]	578	507	551	594	637	680
7.	Tax	45	40	43	47	50	53
8.	**PAT** (6–7)	533	467	508	548	587	626
9.	Add Depreciation	244	244	244	244	244	244
10.	**Net Cash Accruals (Cash Inflow) (8 + 9)**	777	711	752	792	831	871
11.	**Less: Use of Funds (Cash Outflow)**						
	Capital Expenditure (Debt- 1921 + Equity– 2961 = **4882**)						
	Repayment of Term Loan		592	592	592	592	592
	Total Cash Outflow:		592	592	592	592	592
12.	Surplus/Deficit (10–11)	777	119	160	200	239	278

Continue...

		2009	2010	2011	2012	2013	2014
1.	**Total Revenue (without CDM in lacs)**	927	927	927	927	927	927
	CDM Revenue	158	158	158	158	158	158
	Total Revenue (with CDM revenue in lacks)	1085	1085	1085	1085	1085	1085
2.	**Less: Operational Cost**						
	– Operation and Maintenance Cost	100	108	115	124	133	142
	– Insurance	18	18	18	18	18	18
	Total Expenses	119	126	134	142	151	160
3.	PBIDT (1–2)	966	959	951	943	776	766
4.	Depreciation	244	244	244	244	244	244
5.	Interest						
6.	PBT [3–(4 + 5)]	722	715	707	699	532	522
7.	Tax	57	56	55	55	42	41

		2009	2010	2011	2012	2013	2014
8.	**PAT** (6–7)	665	659	652	644	490	481
9.	Add Depreciation	244	244	244	244	244	244
10.	**Net Cash Accruals (Cash Inflow)** (8 + 9)	909	903	896	888	734	725
11.	**Less: Use of Funds (Cash Outflow)**						
	Capital Expenditure (Debt- 1921 + Equity– 2961 = **4882**)						
	Repayment of Term Loan						
	Total Cash Outflow:						
12.	Surplus/Deficit (10-11)	909	903	896	888	734	725
	IRR With CDM Revenue	12.18%					

Source: Researcher's own study.

Without considering the CDM revenues for the above project the IRR works out to 9.49% and with CDM revenues, it works out to 12.18%. The cost of capital for the given capital structure is 10.91% and hence the project is viable only after considering CDM revenues.

Project 2: 1.5 MW Grid connected Wind Electricity Generation at Tirunelveli District, Tamilnadu, India by Kallam Agro Products and Oils Private Limited (2770).

Brief Description

- The project activity is to establish a 1.5 MW Wind Electric Generator (WEG) in Tirunelveli District, Tamilnadu and export the electricity generated to the State grid the Tamil Nadu Electricity Board.

Table 3: Financial Analysis without CDM Revenue (₹ in Lacs)

		2003	2004	2005	2006	2007	2008
1.	**Total Revenue (without CDM)**	24.52	24.52	24.52	24.52	24.52	24.52
2.	**Less: Operational Cost**						
	– Operation and Maintenance Cost	1.07	1.07	1.07	1.07	1.07	1.12
	– Insurance		0.11	0.11	0.11	0.11	0.11
	– Administration and General Overheads	0.30	0.32	0.33	0.35	0.36	0.38
	– Interest on Term Loan	8.26	7.38	6.20	5.02	3.84	2.66
	– Depreciation	4.30	4.30	4.30	4.30	4.30	4.30
	Total Expenses	13.93	13.17	12.01	10.85	9.68	8.57
3.	**PBT** [1–2]	10.59	11.35	12.51	13.67	14.84	15.94
4.	Tax				1.37	1.48	1.59
5.	**PAT** (3–4)	10.59	11.35	12.51	12.31	13.35	14.35

		2003	*2004*	*2005*	*2006*	*2007*	*2008*
6.	Add Depreciation	4.30	4.30	4.30	4.30	4.30	4.30
7.	**Net Cash Accruals (Cash Inflow)** (5 + 6)	14.89	15.65	16.81	16.61	17.65	18.65
	Less: Use of Funds (Cash Outflow)						
	Repayment of Term Loan	5.13	10.26	10.26	10.26	10.26	10.26
	Interest	8.26	7.38	6.20	5.02	3.84	2.66
	Total Cash Outflow:	13.39	17.64	16.46	15.28	14.10	12.92
8.	Surplus/Deficit (10–11)	1.50	(1.99)	0.35	1.33	3.55	5.73

Continue...

		2009	*2010*	*2011*	*2012*	*2013*	*2014*
1.	**Total Revenue (without CDM)**	24.52	24.52	24.52	24.52	24.52	24.52
2.	**Less: Operational Cost**						
	– Operation and Maintenance Cost	1.18	1.24	1.30	1.37	1.44	1.51
	– Insurance	0.11	0.11	0.11	0.11	0.11	0.11
	– Administration and General Overheads	0.40	0.42	0.44	0.47	0.49	0.51
	– Interest on Term Loan	1.48	0.44				
	– Depreciation	4.30	4.30	4.30	4.30	4.30	4.30
	Total Expenses	7.47	6.52	6.16	6.24	6.34	6.43
3.	**PBT** [1–2]	17.05	18	18.36	18.27	18.18	18.09
4.	Tax	1.70	1.80	1.84	1.83	1.82	1.81
5.	**PAT** (6–7)	15.34	16.20	16.53	16.45	16.36	16.28
6.	Add Depreciation	4.30	4.30	4.30	4.30	4.30	4.30
7.	**Net Cash Accruals (Cash Inflow)** (8 + 9)	19.64	20.50	20.83	20.75	20.66	20.58
8.	**Less: Use of Funds (Cash Outflow)**						
	Repayment of Term Loan	10.26	7.70				
	Interest	1.48	0.44				
	Total Cash Outflow:	11.74	8.14				
9.	Surplus/Deficit (10–11)	7.91	12.36	20.83	20.75	20.66	20.58
	IRR Without CDM Revenue	8.35%					

Source: Researcher's own study.

Table 4: Financial Analysis with CDM Revenue (₹ in Lacs)

		2003	2004	2005	2006	2007	2008
1.	Total Revenue (without CDM in lacs)	24.52	24.52	24.52	24.52	24.52	24.52
2.	CDM Revenue	5.70	5.70	5.70	5.70	5.70	5.70
3.	**Total Revenue (with CDM revenue in lacs)**	30.21	30.21	30.21	30.21	30.21	30.21
4.	**Less: Operational Cost**						
	– Operation and Maintenance Cost	1.07	1.07	1.07	1.07	1.07	1.12
	– Insurance		0.11	0.11	0.11	0.11	0.11
	– Administration and General Overheads	0.30	0.32	0.33	0.35	0.36	0.38
	– Interest on Term Loan	8.26	7.38	6.20	5.02	3.84	2.66
	– Depreciation	4.30	4.30	4.30	4.30	4.30	4.30
	Total Expenses	13.93	13.17	12.01	10.85	9.68	8.57
5.	**PBT [3–4]**	16.28	17.04	18.20	19.37	20.53	21.64
6.	Tax	1.30	1.36	1.46	1.55	1.64	1.73
7.	**PAT (5–6)**	14.98	15.68	16.75	17.82	18.89	19.91
8.	Add Depreciation	4.30	4.30	4.30	4.30	4.30	4.30
9.	**Net Cash Accruals (Cash Inflow) (7 + 8)**	19.28	19.98	21.05	22.12	23.19	24.21
10.	**Less: Use of Funds (Cash Outflow)**						
	Repayment of Term Loan	5.13	10.26	10.26	10.26	10.26	10.26
	Interest	8.26	7.38	6.20	5.02	3.84	2.66
	Total Cash Outflow:	13.39	17.64	16.46	15.28	14.10	12.92
11.	Surplus/Deficit (9–10)	5.89	2.34	4.59	6.84	9.09	11.29

Continue...

		2009	2010	2011	2012	2013	2014
1.	Total Revenue (without CDM in lacs)	24.52	24.52	24.52	24.52	24.52	24.52
2.	CDM Revenue	5.70	5.70	5.70	5.70	5.70	5.70
3.	**Total Revenue (with CDM revenue in lacs)**	30.21	30.21	30.21	30.21	30.21	30.21
4.	**Less: Operational Cost**						
	– Operation and Maintenance Cost	1.18	1.24	1.30	1.37	1.44	1.51
	– Insurance	0.11	0.11	0.11	0.11	0.11	0.11
	– Administration and General Overheads	0.40	0.42	0.44	0.47	0.49	0.51

		2009	*2010*	*2011*	*2012*	*2013*	*2014*
	– Interest on Term Loan	1.48	0.44				
	– Depreciation	4.30	4.30	4.30	4.30	4.30	4.30
	Total Expenses	7.47	6.52	6.16	6.24	6.34	6.43
5.	**PBT [3–4]**	22.74	23.70	24.06	23.97	23.88	23.78
6.	Tax	1.82	1.90	1.92	1.92	1.91	1.90
7.	**PAT (5–6)**	20.92	21.80	22.13	22.05	21.97	21.88
8.	Add Depreciation	4.30	4.30	4.30	4.30	4.30	4.30
9.	**Net Cash Accruals (Cash Inflow)** (7 + 8)	25.22	26.10	26.43	26.35	26.27	26.18
10.	**Less: Use of Funds (Cash Outflow)**						
	Repayment of Term Loan	10.26	7.70				
	Interest	1.48	0.44				
	Total Cash Outflow:	11.74	8.14				
11.	Surplus/Deficit (9–10)	13.49	17.96	26.43	26.35	26.27	26.18
	IRR with CDM Revenue	12.75%					

Source: Researcher's own study.

Without considering the CDM revenues the IRR works out to 8.35% and with CDM revenues, it works out to 12.75%. The cost of capital for the given capital structure is 11.85% and hence the project is viable only after considering CDM revenues.

Project 3: Wind Power based electricity generation project in India by DLF Home Developers Limited (3642).

Brief Description
- The project activity involves the establishment of a wind farm of 67.5 MW capacity enabling generation of electricity by state-of-art 1.5 MW capacity Wind Energy Generators (WEGs).

Table 5: Financial Analysis without CDM Revenue (₹ in Lacs)

		2003	*2004*	*2005*	*2006*	*2007*	*2008*
1.	**Total Revenue (without CDM)**	284.1	284.1	284.1	284.1	284.1	284.1
2.	**Less: Operational Cost**						
	– Operation and Maintenance Cost	16.8	17.7	18.5	19.5	20.4	21.5
	– Interest on Term Loan	88.3	83.8	75	66.2	57.4	48.5
	– Depreciation	45.1	45.1	45.1	45.1	45.1	45.1
	Total Expenses	150.2	146.6	138.7	130.8	122.9	115.1
3.	**PBT [1–2]**	134	137.5	145.5	153.4	161.2	169
4.	Tax						16.9

		2003	2004	2005	2006	2007	2008
5.	**PAT** (3–4)	134	137.5	145.5	153.4	161.2	152.1
6.	Add Depreciation	45.1	45.1	45.1	45.1	45.1	45.1
7.	**Net Cash Accruals (Cash Inflow)** (5 + 6)	179.1	182.6	190.6	198.5	206.3	197.2
8.	**Less: Use of Funds (Cash Outflow)**						
	Repayment of Term Loan		84.1	84.1	84.1	84.1	84.1
	Interest	88.3	83.8	75	66.2	57.4	48.5
	Total Cash Outflow:	88.3	167.9	159.1	150.3	141.4	132.6
9.	Surplus/Deficit (7–8)	90.8	14.7	31.5	48.2	64.9	64.6

Continue…

		2009	2010	2011	2012	2013	2014
1.	**Total Revenue (without CDM)**	284.1	284.1	284.1	284.1	284.1	284.1
2.	**Less: Operational Cost**						
	– Operation and Maintenance Cost	22.5	23.7	24.8	26.1	26.1	26.1
	– Interest on Term Loan	39.7	30.9	22.1	13.2	4.4	39.7
	– Depreciation	45.1	45.1	45.1	45.1	45.1	45.1
	Total Expenses	107.3	99.6	92	84.4	75.6	71.2
3.	**PBT** [1–2]	176.8	184.5	192.1	199.7	208.5	213
4.	Tax	17.7	18.4	19.2	20	20.9	21.3
5.	**PAT** (3–4)	159.1	166	172.9	179.7	187.7	191.7
6.	Add Depreciation	45.1	45.1	45.1	45.1	45.1	45.1
7.	**Net Cash Accruals (Cash Inflow)** (5 + 6)	204.2	211.1	218	224.8	232.8	236.8
8.	**Less: Use of Funds (Cash Outflow)**						
	Repayment of Term Loan	84.1	84.1	84.1	84.1	84.1	
	Interest	39.7	30.9	22.1	13.2	4.4	
	Total Cash Outflow:	123.8	115	106.1	97.3	88.5	
9.	Surplus/Deficit (7–8)	80.4	96.2	111.9	127.5	144.3	236.8
	IRR without CDM Revenue	10.41%					

Source: Researcher's own study.

Table 6: Financial Analysis with CDM Revenue (₹ in Lacs)

		2003	2004	2005	2006	2007	2008
1.	Total Revenue (without CDM in lacs)	284.1	284.1	284.1	284.1	284.1	284.1
2.	CDM Revenue	52.5	52.5	52.5	52.5	52.5	52.5
3.	**Total Revenue (with CDM revenue in lacs)**	336.6	336.6	336.6	336.6	336.6	336.6
4.	**Less: Operational Cost**						
	– Operation and Maintenance Cost	16.8	17.7	18.5	19.5	20.4	21.5

		2003	2004	2005	2006	2007	2008
	– Interest on Term Loan	88.3	83.8	75	66.2	57.4	48.5
	– Depreciation	45.1	45.1	45.1	45.1	45.1	45.1
	Total Expenses	150.2	146.6	138.7	130.8	122.9	115.1
5.	**PBT** [1–2]	186.5	190	198	205.9	213.7	221.5
6.	Tax						22.2
7.	**PAT** (3–4)	186.5	190	198	205.9	213.7	199.4
8.	Add Depreciation	45.1	45.1	45.1	45.1	45.1	45.1
9.	**Net Cash Accruals (Cash Inflow)** (5 + 6)	231.6	235.1	243.1	251	258.8	244.5
10.	**Less: Use of Funds (Cash Outflow)**						
	Repayment of Term Loan		84.1	84.1	84.1	84.1	84.1
	Interest	88.3	83.8	75	66.2	57.4	48.5
	Total Cash Outflow:	88.3	167.9	159.1	150.3	141.4	132.6
11.	Surplus/Deficit (9–10)	143.3	67.2	84	100.7	117.4	111.

Continue...

		2009	2010	2011	2012	2013	2014
1.	Total Revenue (without CDM in lacs)	284.1	284.1	284.1	284.1	284.1	284.1
2.	CDM Revenue	52.5					
3.	**Total Revenue (with CDM revenue in lacs)**	336.6	284.1	284.1	284.1	284.1	284.1
4.	**Less: Operational Cost**						
	– Operation and Maintenance Cost	39.7	30.9	22.1	13.2	4.4	
	– Interest on Term Loan	22.5	23.7	24.8	26.1	26.1	26.1
	– Depreciation	45.1	45.1	45.1	45.1	45.1	45.1
	Total Expenses	107.3	99.6	92	84.4	75.6	71.2
5.	**PBT** [1–2]	229.3	184.5	192.1	199.7	208.5	213
6.	Tax	22.9	18.4	19.2	20	20.9	21.3
7.	**PAT** (3–4)	206.4	166	172.9	179.7	187.7	191.7
8.	Add Depreciation	45.1	45.1	45.1	45.1	45.1	45.1
9.	**Net Cash Accruals (Cash Inflow)** (5 + 6)	251.5	211.1	218	224.8	232.8	236.8
10.	**Less: Use of Funds (Cash Outflow)**						
	Repayment of Term Loan	84.1	84.1	84.1	84.1	84.1	
	Interest	39.7	30.9	22.1	13.2	4.4	
	Total Cash Outflow:	123.8	115	106.1	97.3	88.5	
11.	Surplus/Deficit (9–10)	127.7	96.2	111.9	127.5	144.3	236.8
	IRR with CDM Revenue	13.21%					

Source: Researcher's own study.

Without considering the CDM revenues the IRR works out to 10.41% and with CDM revenues, it works out to 13.21%. The cost of capital for the given capital structure is 11.38% and hence the project is viable only after considering CDM revenues.

Project 4: 13.75 MW wind power project in Davangere, Karnataka, India. (3700)

Brief Description

- The project activity is an initiative by Sargam Retails Private Limited (SRPL), towards clean electricity generation using wind energy resources in the state of Karnataka. SRPL is engaged in Trading and Marketing Packaged Tea and Tobacco.

Table 7: Financial Analysis without CDM Revenue (₹ in Lacs)

		2003	*2004*	*2005*	*2006*	*2007*	*2008*
1.	**Total Revenue (without CDM)**	2100	2100	2100	2100	2100	2100
2.	**Less: Operational Cost**						
	– Operation and Maintenance Cost		130	137	143	150	158
	– Service Tax on O and M @12.36%		16	17	18	19	20
	– Insurance	21	21	21	21	21	21
	– Interest	678	621	508	395	282	169
	– Depreciation	406	406	406	406	406	406
	Total Expenses	1106	1195	1090	984	879	775
3.	**PBT** [1–2]	994	905	1010	1116	1221	1325
4.	Tax				112	122	133
5.	**PAT** (3–4)	994	905	1010	1004	1099	1193
6.	Add Depreciation	406	406	406	406	406	406
7.	**Net Cash Accruals (Cash Inflow)** (5 + 6)	1400	1311	1417	1410	1505	1599
8.	**Less: Use of Funds (Cash Outflow)**						
	Repayment of Term Loan		962	962	962	962	962
	Interest	678	621	508	395	282	169
	Total Cash Outflow:	678	1583	1470	1357	1244	1131
9.	Surplus/Deficit (7–8)	723	(272)	(53)	53	261	468

Continue...

		2009	*2010*	*2011*	*2012*	*2013*	*2014*
1.	**Total Revenue (without CDM)**	2100	2100	2100	2100	2100	2100
2.	**Less: Operational Cost**						
	– Operation and Maintenance Cost	166	174	183	192	202	212
	– Service Tax on O and M @12.36%	21	22	23	24	25	26
	– Insurance	21	21	21	21	21	21

		2009	*2010*	*2011*	*2012*	*2013*	*2014*
	− Interest	56					
	− Depreciation	406	406	406	406	406	406
	Total Expenses	671	623	633	644	654	666
3.	**PBT** [1–2]	1429	1477	1467	1456	1446	1434
4.	Tax	143	148	147	146	145	143
5.	**PAT** (3–4)	1286	1329	1320	1311	1301	1291
6.	Add Depreciation	406	406	406	406	406	406
7.	**Net Cash Accruals (Cash Inflow)** (5 + 6)	1693	1735	1726	1717	1707	1697
8.	**Less: Use of Funds (Cash Outflow)**						
	Repayment of Term Loan	962					
	Interest	56					
	Total Cash Outflow:	1018					
9.	Surplus/Deficit (7–8)	675	1735	1726	1717	1707	1697
	IRR without CDM Revenue	9.34%					

Source: Researcher's own study.

Table 8: Financial Analysis with CDM Revenue (₹ in Lacs)

		2003	*2004*	*2005*	*2006*	*2007*	*2008*
1.	Total Revenue (without CDM in lacs)	2100	2100	2100	2100	2100	2100
2.	CDM Revenue	487	487	487	487	487	487
3.	**Total Revenue (with CDM revenue in lacs)**	2587	2587	2587	2587	2587	2587
4.	**Less: Operational Cost**						
	− Operation and Maintenance Cost		130	137	143	150	158
	− Service Tax on O and M @12.36%		16	17	18	19	20
	− Insurance	21	21	21	21	21	21
	− Interest	678	621	508	395	282	169
	− Depreciation	406	406	406	406	406	406
	Total Expenses	1106	1195	1090	984	879	775
5.	**PBT** [3–4]	1481	1391	1497	1602	1707	1812
6.	Tax	163	153	165	176	188	199
7.	**PAT** (5–6)	1318	1238	1332	1426	1520	1613
8.	+ Depreciation	163	153	165	176	188	199
9.	**Net Cash Accruals (Cash inflow)** (7 + 8)	1724	1645	1739	1832	1926	2019
10.	**Less: Use of Funds (Cash Outflow)**						
	Repayment of Term Loan		962	962	962	962	962
	Interest	678	621	508	395	282	169
	Total Cash Outflow:	678	1583	1470	1357	1244	1131
11.	Surplus/Deficit (9–10)	1046	62	269	475	682	88

Continue....

		2009	2010	2011	2012	2013	2014
1.	Total Revenue (without CDM in lacs)	2100	2100	2100	2100	2100	2100
2.	CDM Revenue	487	487	487	487	487	487
3.	**Total Revenue (with CDM revenue in lacs)**	2587	2587	2587	2587	2587	2587
4.	**Less: Operational Cost**						
	– Operation and Maintenance Cost	166	174	183	192	202	212
	– Service Tax on O and M @12.36%	21	22	23	24	25	26
	– Insurance	21	21	21	21	21	21
	– Interest	56					
	– Depreciation	406	406	406	406	406	406
	Total Expenses	671	623	633	644	654	666
5.	**PBT [3–4]**	1916	1963	1953	1943	1446	1434
6.	Tax	211	216	215	214	159	158
7.	**PAT (5–6)**	1705	1747	1739	1729	1287	1277
8.	+ Depreciation	211	216	215	214	159	158
9.	**Net Cash Accruals (Cash inflow) (7 + 8)**	2112	2154	2145	2136	1693	1683
10.	**Less: Use of Funds (Cash Outflow)**						
	Repayment of Term Loan						
	Interest						
	Total Cash Outflow:						
11.	Surplus/Deficit (9–10)	2154	2145	2136	1693	1683	2154
	IRR With CDM Revenue	12.43%					

Source: Researcher's own study.

Without considering the CDM revenues the IRR works out to 9.34% and with CDM revenues, it works out to 12.43%. The cost of capital for the given capital structure is 11.82% and hence the project is viable only after considering CDM revenues.

Project 5: MW Wind-Power Project in Tamil Nadu by REI Agro Limited (3710).

Brief Description

- The project activity consists of 12 nos. Vestas RRB make 500 KW wind turbines totaling to a capacity of 6 MW.

Table 9: Financial Analysis without CDM Revenue (₹ in Lacs)

		2003	*2004*	*2005*	*2006*	*2007*	*2008*
1.	**Total Revenue (without CDM)**	105.1	105.1	105.1	105.1	105.1	105.1
2.	**Less: Operational Cost**						
	– Operation and Maintenance Cost	9.3	9.7	10.2	10.7	11.3	11.8
	– Insurance	2.8	2.8	2.8	2.8	2.8	2.8
	Total Expenses	12	12.5	13	13.5	14	14.6
3.	**PBIDT** (1–2)	93.1	92.6	92.1	91.6	91.1	90.5
4.	Interest	29.3	27.7	24.4	21.2	17.9	14.6
5.	Depreciation	18.6	18.6	18.6	18.6	18.6	18.6
6.	**PBT** [3–(4 + 5)]	45.2	46.3	49.1	51.9	54.6	57.3
7.	Tax				5.2	5.5	5.7
8.	**PAT** (6–7)	45.2	46.3	49.1	46.7	49.1	51.5
9.	+ Depreciation	18.6	18.6	18.6	18.6	18.6	18.6
10.	**Net Cash Accruals (Cash Inflow)** (8 + 9)	63.8	64.9	67.7	65.3	67.7	70.1
11.	**Less: Use of Funds (Cash Outflow)**						
	Repayment of Term Loan		31	31	31	31	31
	Interest	29.3	27.7	24.4	21.2	17.9	14.6
	Total Outflow:	29.3	58.7	55.4	52.2	48.9	45.6
12.	Surplus/Deficit (10–11)	34.5	6.3	12.3	13.1	18.8	24.5

		2009	*2010*	*2011*	*2012*	*2013*	*2014*
1.	**Total Revenue (without CDM)**	105.1	105.1	105.1	105.1	105.1	105.1
2.	**Less: Operational Cost**						
	– Operation and Maintenance Cost	12.4	13	13.7	14.4.	15.1	15.8
	– Insurance	2.8	2.8	2.8	2.8	2.8	2.8
	Total Expenses	15.2	15.8	16.5	17.2	17.9	18.6
3.	**PBIDT** (1–2)	89.9	89.3	88.7	88	87.3	86.5
4.	Interest	11.4	8.1	4.9	1.6		
5.	Depreciation	18.6	18.6	18.6	18.6	18.6	18.6
6.	**PBT** [3–(4 + 5)]	59.9	62.6	65.2	67.7	68.7	67.9

		2009	*2010*	*2011*	*2012*	*2013*	*2014*
7.	Tax	6	12.5	13	13.5	13.7	13.6
8.	**PAT** (6–7)	53.9	50.1	52.1	54.2	54.9	54.3
9.	+ Depreciation	18.6	18.6	18.6	18.6	18.6	18.6
10.	**Net Cash Accruals (Cash Inflow)** (8 + 9)	72.5	68.7	70.7	72.8	73.5	72.9
11.	**Less: Use of Funds (Cash Outflow)**						
	Repayment of Term Loan	31	31	31	31		
	Interest	11.4	8.1	4.9	1.6		
	Total Outflow:	42.4	39.1	35.9	32.6		
12.	Surplus/Deficit (10–11)	30.1	29.5	34.9	40.2	73.5	72.9
	IRR Without CDM Revenue	10.36%					

Source: Researcher's own study.

Table 10: Financial Analysis with CDM Revenue (₹ in Lacs)

		2003	*2004*	*2005*	*2006*	*2007*	*2008*
1.	Total Revenue (without CDM in lacs)	105.1	105.1	105.1	105.1	105.1	105.1
2.	CDM Revenue	16.4	16.4	16.4	16.4	16.4	16.4
3.	**Total Revenue (with CDM revenue in lacs)**	121.5	121.5	121.5	121.5	121.5	121.5
4.	**Less: Operational Cost**						
	– Operation and Maintenance Cost	109.5	109	108.5	108	107.5	106.9
	– Insurance	29.3	27.7	24.4	21.2	17.9	14.6
	Total Expenses	18.6	18.6	18.6	18.6	18.6	18.6
	PBIDT (4–5)	61.6	62.7	65.5	68.2	71	73.7
	Interest				6.8	7.1	7.4
	Depreciation	61.6	62.7	65.5	61.4	63.9	66.3
5.	**PBT** [3–4]	18.6	18.6	18.6	18.6	18.6	18.6
6.	Tax	80.2	81.3	84.1	80	82.5	84.9
7.	**PAT** (5–6)	109.5	109	108.5	108	107.5	106.9
8.	+ Depreciation	29.3	27.7	24.4	21.2	17.9	14.6
9.	**Net Cash Accruals (Cash Inflow)** (7 + 8)	18.6	18.6	18.6	18.6	18.6	18.6
10.	**Less: Use of Funds (Cash Outflow)**						
	Repayment of Term Loan		31	31	31	31	31
	Interest	29.3	27.7	24.4	21.2	17.9	14.6
	Total Cash Outflow:	29.3	58.7	55.4	52.2	48.9	45.6
11.	Surplus/Deficit (9–10)	50.9	22.7	28.7	27.9	33.6	39.2

		2009	2010	2011	2012	2013	2014
1.	Total Revenue (without CDM in lacs)	105.1	105.1	105.1	105.1	105.1	105.1
2.	CDM Revenue	16.4	16.4	16.4	16.4		
3.	**Total Revenue (with CDM revenue in lacs)**	121.5	121.5	121.5	121.5	105.1	105.1
4.	**Less: Operational Cost**						
	– Operation and Maintenance Cost	12.4	13	13.7	14.4.	15.1	15.8
	– Insurance	2.8	2.8	2.8	2.8	2.8	2.8
	Total Expenses	15.2	15.8	16.5	17.2	17.9	18.6
	PBIDT (4–5)	106.3	105.7	105	104.4	87.3	86.5
	Interest	11.4	8.1	4.9	1.6		
	Depreciation	18.6	18.6	18.6	18.6	18.6	18.6
5.	**PBT** [3–4]	76.3	79	81.6	84.1	68.7	67.9
6.	Tax	7.6	15.8	16.3	16.8	13.7	13.6
7.	**PAT** (5–6)	68.7	63.2	65.2	67.3	54.9	54.3
8.	+ Depreciation	18.6	18.6	18.6	18.6	18.6	18.6
9.	**Net Cash Accruals (Cash Inflow) (7 + 8)**	87.3	81.8	83.8	85.9	73.5	72.9
10.	**Less: Use of Funds (Cash Outflow)**						
	Repayment of Term Loan	31	31	31	31		
	Interest	11.4	8.1	4.9	1.6		
	Total Cash Outflow:	42.4	39.1	35.9	32.6		
11.	Surplus/Deficit (9–10)	44.9	42.6	48	53.3	73.5	72.9
	IRR With CDM Revenue	13.89%					

Source: Researcher's own study.

Without considering the CDM revenues the IRR works out to 10.36% and with CDM revenues, it works out to 13.89%. The cost of capital for the given capital structure is 11.38% and hence the project is viable only after considering CDM revenues.

4. GROWTH IN CDM SECTOR LENDING

A few years back the development finance institutions (DFIs) were engaged in pure lending for solid ventures. Today only few of the long term lending institutions like, PFC and IREDA are engaged in this activity. However, the majority of the renewable energy based projects that have been cleared in India have been able to get loans from these institutions. This is mainly due to the mandates and subsidies given by the Government to the above FIs promoted by it. The year wise and sector wise sanctions of loans to CDM projects in put forth.

5. LENDING BY IREDA

Table 11: Year-wise Sanctions and Disbursements
by IREDA to CDM projects in India (₹ in Crores)

Year	Total Sanctions	Total Disbursements
2003–04	423.57	343.28
2004–05	599.73	289.98
2005–06	505.83	302.51
2006–07	588.51	410.87
2007–08	826.15	553.64
2008–09	1489.93	770.95
2009–10	1823.91	890.03
2010–11	3126.42	1224.17
2011–12	3405.96	1855.04
2012–13	3747.36	2125.5

Table 12: Sector Wise Sanctions of IREDA's Loans to CDM Projects in India (₹ in Crores)

	Wind Power	Hydro Power	Co-generation	Biomass Power	Energy Efficiency and Conservation	Solar Photo Volataic	Solar Thermal
2003–04	121.05	122	58.56	58.29	45.03	2.96	11.43
2004–05	204.25	176.73	77.46	32.2	96.73	0	12.36
2005–06	261.41	17.3	0	89.3	123.32	0	7
2006–07	266.19	160.87	116.28	0	21.3	0	13
2007–08	426.97	226.23	68.3	0	53.73	0	50.92
2008–09	728.87	343.4	319.85	16.25	40.2	33.36	8
2009–10	1174.1	483.45	140.12	17.25	0	0	0
2010–11	1495.6	984.51	279.15	49.2	261.56	39.39	0
2011–12	1643.5	772.93	499.65	0	141.13	344.81	0
2012–13	1792.2	914.46	711.12	0	0	321.51	0

Continue...

	Waste to Energy	Biomethanation from Industrial Effluents	Biomass Briquetting	Biomass Gasification	Miscellaneous	Total
2003–04	0	0	0	0	4.25	423.57
2004–05	0	0	0	0	0	599.73
2005–06	0	0	0	0	7.5	505.83
2006–07	9.19	0	1.68	0	0	588.51
2007–08	0	0	0	0	0	826.15
2008–09	0	0	0	0	0	1489.93
2009–10	0	0	0	0	0	1823.91
2010–11	16.98	0	0	0	0	3126.42
2011–12	3.9	0	0	0	0	3405.96
2012–13	8.1	0	0	0	0	3747.36

Table 13

	Wind Power	Hydro Power	Co-generation	Biomass Power	Energy Efficiency and Conservation	Solar Photo Volataic	Solar Thermal
2003–04	93.37	79.52	86.03	52.69	5.3	6.97	7.87
2004–05	102.47	90.04	26.57	45.67	0.65	5.57	17.25
2005–06	134.82	64.73	18.44	36.74	41.44	0.11	5.93
2006–07	258.19	58.36	19.68	38.99	29.4	0	5
2007–08	271.02	119.39	103.88	9.81	13.57	0	29.82
2008–09	483.51	147.55	76.36	1.13	5.8	26.25	27.55
2009–10	515.92	229.03	83.49	24.37	15.18	7.06	14.51
2010–11	644.34	340.49	180.47	45.81	8	0	0
2011–12	1199.5	165.76	330.92	0	73.25	83.47	0
2012–13	1207.9	356.28	347.94	0	59.74	151.2	0

Sector Wise disbursement of IREDA's Loans to CDM Projects in India (₹ in Crores).

Continue...

	Waste to Energy	Biomethanation from Industrial Effluents	Biomass Briquetting	Biomass Gasification	Miscellaneous	Total
2003–04	9.52	0	0.07	0	1.94	343.28
2004–05	1.64	0	0	0	0.12	289.98
2005–06	0.3	0	0	0	0	302.51
2006–07	0	0	0	1.25	0	410.87
2007–08	6.15	0	0	0	0	553.64
2008–09	2.8	0	0	0	0	770.95
2009–10	0.47	0	0	0	0	890.03
2010–11	5.06	0	0	0	0	1224.17
2011–12	2.13	0	0	0	0	1855.03
2012–13	2.45	0	0	0	0	2125.5

Table 14: Sanction of Loans by IREDA to CDM Projects in Maharashtra (₹ in Crores)

Year	IREDA's Loan Amount Sanctioned in Maharashtra	IREDA's Total Sanctions in India	% Share of Loan Amount Sanctioned in Maharashtra to Total Sanctions in India
2003–04	70.61	423.57	16.67
2004–05	54.82	599.73	9.14
2005–06	139.52	505.83	27.58
2006–07	215.31	588.51	36.59
2007–08	35.65	826.15	4.32
2008–09	100.25	1489.93	6.73
2009–10	189.8	1823.91	10.41
2010–11	51.27	3126.42	1.64
2011–12	521.05	3405.96	15.3
2012–13	945.25	3747.36	25.22

Table 15: Disbursement of Loans by IREDA to CDM Projects in Maharashtra (₹ in Crores)

Year	IREDA's Loan Amount Sanctioned in Maharashtra	IREDA's Total Sanctions in India	% Share of Loan Amount Sanctioned in Maharashtra to Total Sanctions in India
2003–04	42.11	343.28	12.27
2004–05	12.35	289.98	4.26
2005–06	26.04	302.51	8.61
2006–07	162.45	410.87	39.54
2007–08	137.41	553.64	24.82
2008–09	15.49	770.95	2.01
2009–10	112.85	890.03	12.68
2010–11	330.3	1224.17	26.98
2011–12	393.35	1855.03	21.20
2012–13	437.31	2125.5	20.57

6. CALCULATION OF CER ISSUED FROM CDM PROJECTS

Null Hypothesis (H0): The Certified Emission Reductions (CER) issued by CDM projects in India are growing

Alternative Hypothesis (H1): The Certified Emission Reductions (CER) issued by CDM projects in India are not growing.

Table 16: Year-wise Issuance of CER

Year of Issuance	Total CER Issued
2005	48230
2006	12950789
2007	21066026
2008	20117353
2009	18727157
2010	9414269
2011	44791261
2012	35423765
2013	26441099
2014	11264557
2015	2583966

Graph 1 Growth of Total Expected Accumulated 2012 CERs

http://www.cdmpipeline.org/cers.htm

Responses collected through questionnaires with respect to barriers and challenges confronted from selected corporate was analyzed and the following challenges were majorly put forward by many of the respondents:

(a) *Project Developers' Perspective*

- Potential project developers are not aware that their projects can generate emission credits.
- Up-and down enthusiasm on CDM (KP Ratification in 2004 and DNA in 2005.
- Technical issues: CDM project development involves jargons and conditions (such as additionality, baseline scenario, boundaries) which are not practical and time consuming to project developers.
- Although consolidated methodologies are built to address this technical issues, data needs may not be easily available.
- Weak understanding of CDM modalities and procedures.
- Little information on and contacts with emission credit buyers.

(b) *Lack of "CDM Awareness" of Related Institution that May Help the Project Developers*

- The institutions supposed to support the project developers (financial institutions, local consultants, local designated operational entities, business associations, local media/press) are not aware of CDM and its opportunities/advantages.
- CDM provides incentives but those who are aware do not have clear understanding on CDM t project financing
- "CDM Brokers"

(c) *Risks and Uncertainties*

- Complicated process and continuous development of the CDM procedures, and unavailable support from local institutions (intermittent awareness and effort).
- Obtaining PPA in power projects to secure underlying financing Some project risks undermine CDM attractiveness.
- CER ownership: in some sector CER ownership will be an issue due to national policy/regulation.

7. CONCLUSION

Based on the data analysis, it was found that the IRR of the companies associated with CDM increases with the CDM revenue and vice versa. There overall trend of lending to CDM projects was growing with the passage of time. The CER issued to CDM projects was rising which was also very clear with the growth in the registered CDM projects in India. India stands next to china with respect to the number of registered CDM projects.

REFERENCES

Bakker (2009). Differentiation in the CDM options and impacts.Scientific Assessment and Policy Analysis. Energy research Centre of the Netherlands (ECN).

Bohringer (2005). The Kyoto Protocol: Success or Failure, in Helm,Climate Change Policy Oxford, UK: Oxford University Press.

Brown (2004). How do CDM projects contribute to sustainable development? Tyndall Centre for Climate Change Research Technical Report 16.

Cames (2003). CDM Host Country Institution Building, Mitigation and Adaptation. Strategies for Global Change 8: 201–220.

Chung (2007). A CER Discounting Scheme could Save Climate Change Regime after 2012. Climate Policy 7, 171–176.

Cosbey (2007). Market Mechanisms for Sustainable Development: How Do They Fit in the Various Post–2012 Climate Efforts. The Development Dividend Project, phase III. International Institute for Sustainable Development (IISD).

Curnow (2009). Implementing CDM–A guidebook to host country legal issues. Baker and McKenzie and UNEP Risoe Centre.

Ghosh (2007). The Clean Development Mechanism and some open questions. Oslo: Development Today.

Grubb (1999). The Kyoto Protocol: A Guide and Assessment. . RIIA, Earthscan, 256–268.

Haya (2009). Measuring Emissions against an Alternative Future: Fundamental Flaws in the Structure of the Kyoto Protocol's Clean Development Mechanism. Energy and Resources Group Working Paper ERG09–001. University of California, Berkeley.

Kalia (2011). Sustainability—The Way Ahead for Banking and Financial Institutions. Ernst and Young.

Kamel (2007). Capacity Development for CDM Project: Guidebook to financing CDM Projects. UNEP, Ecosecurities.

Kolshus (2001). Can the Clean Development Mechanism Attain Both Cost-effectiveness and Sustainable Development Objectives? CICERO Working Paper.

Lohmann (2006). Carbon trading–A critical conversation on climate change, Privatization and power. Development Dialogue Dag Hammarskjöld Foundation, Corner House.

Lloyd's (2006). Climate change–adapt or bust. 360 Risk Project Catastrophy Trends.

Matschoss (2007). The Programmatic Approach to CDM: Benefits for Energy Efficiency Projects. CCLR The Carbon and Climate Law Review, Volume 1 (2):, Lexxion, 119–128.

Mendis (2004). The clean development mechanism: making it operational. Environment, Development and Sustainability 6(1), 183–211.

Michaelowa (2008). Empirical Analysis of Performance of CDM Projects, Political Economy and development Climate policy.

Michaelowa (2005). Transaction costs, institutional rigidities and the size of the Clean Development Mechanism. Energy Policy 33, 511–523.

Michaelowa (2003). Transaction costs of the Kyoto Mechanisms. Climate Policy 3, 261–278.

Michealowa (2007). Additionality determination of Indian CDM projects. Can Indian CDM project developers outwit the CDM Executive Board? Climate Strategies.

Nagai (2005). How Cost-effective are Carbon Emission Reduction under the Prototype Carbon Fund? Environmental Change Institute, University of Oxford.

Olsen (2008). Sustainable Development Benefits of Clean Development Mechanism Projects. A New Methodology for Sustainability Assessment based on Text Analysis of the Project Design Documents Submitted for Validation. Energy policy 36, 2819–2830.

Pearce (2005). The Social Cost of Carbon, in Helm. Climate Change Policy UK: Oxford University Press, chapter 5.

Sirohi (2007). CDM: Is it a 'win–win' strategy for rural poverty alleviation in India? Climatic Change, 91–110.

Schatz (2008). Discounting the Clean Development Mechanism. Georgetown International Environmental Law Review 20 (4), 704–742.

Schneider (2007). Is the CDM Fulfilling its environmental and sustainable development objectives? An evaluation of the CDM and options for improvement. Climate Strategies.

Sterk (2006). Enhancing the Clean Development Mechanism through Sectoral Approach Definitions, Applications and Ways Forward. International Environmental Agreement: Politics, Law and Economics, 271–287.

Streck (2007). Making Markets Works: A Review of CDM Performance and the Need for Reform. European Journal of International Law 19 (2), 409–442.

Sutter (2007). Does the current Clean Development Mechanism (CDM) deliver its sustainable development claim? An analysis of officially registered CDM projects. Climatic Change, 75–90.

Thorne (2008). Towards a framework of clean energy technology receptivity. Energy Policy, 36 (8), pp. 2821–2828.

UNEP Risoe (2010). UNEP Risoe CDM/JI Pipeline Analysis and Database. http://cdmpipeline.org/ www.cd4cdm.org and www.uneprisoe.org.

Victor (2008). A Realistic Policy on International Carbon Offsets. PESD Working Paper 74.

Wara (2007). Is the Global Carbon Market Working? Nature 445(8), 595–96.

Zegras (2007). As if Kyoto mattered: the Clean Development Mechanism and Transportation. Energy Policy 35: 5136–5050.

Logistic Optimisation—An Approach Towards Sustainable Supply Chains

Jyoti Behera Shuvra[1], Anupama Panghal[2] and Swati Baronia[3]

[1,2] Department of Food Business Management, NIFTEM, MOFPI,
Govt. of India, Kundli, India

[3] Food Supply Chain Management, NIFTEM, MOFPI, Govt. of India, Kundli, India

E-mail: [1]harsharyaan@gmail.com; [2]anupamakhatkar@gmail.com;
[3]swatibaronia@gmail.com

ABSTRACT

- *Aim: The aim of this research paper is to outline the hypothesis that will encompass the role of optimized supply chain distribution network towards achieving the sustainable and green supply chain.*

- *Methodology: The methodology will describe firstly, to identify the optimised transportation route for the movement of the finished goods. Secondly, to locate the right position of the warehouse or distribution centres in a supply chain. And thirdly decreasing the overall freight cost incurred during movement of goods from factory to warehouse and hub to warehouse in a logistic operation. The paper is based on theoretical approach towards proposing a model for acheiving a sustainable supply chain.*

- *Findings: The paper highlights the need for effective implementation of proposed strategy or model focusing on logistic optimization, to achieve a sustainable supply chain. Implementing GSCM (green supply chain management) in terms of logistic optimization will ensure the enterprises to achieve growth in market share and profitability by reducing the environmental risks such as carbon accumulation in the food products and carbon emission to the environment. This philosophy is not only responsible for improved service levels and delivery time of finished goods but also responsible for proper channelization of energy such that the emission of carbon to environment can be reduced. The proposed network optimization model helps in locating the position of the warehouse or distribution centres along the supply chain in such a way that the number of kilometres that the vehicle travels will reduce. This in turn will be the prime factor for sustainable supply chain to utilize less energy while movement and lesser carbon footprint in the environment through the different nodes of supply chain.*

- *Managerial Implications: The innovated strategist model will not only be the solution to reduce the carbon foot print in the supply chain stages but also to know where there is a requirement to develop our understanding of the past performance measurements and establish the relationships of the basic types of measurements to the key logistic processes in supply chain. Traditionally the supply chain managers were too much focussed on planning right inventory policies, transport selection and routing and facility location. But now the focus on all these activities is with the objective of linking them with the sustainable supply chain to increase revenue and high return on investments. The proposed methodology has been tested by getting it adopted in one of the FMCG Company in India to improve the profitability and match the demand and supply of goods for consumer in order to make sustainable supply chain.*

- *Originality: The paper has adopted a unique approach of integrating logistics optimisation technique with the concept of reduced carbon footpinting levels in the supply chain. The suggested model for Optimisation Strategies for acheiving Sustaianabilty in Supply Chain has been conceptualised with a real life example.*
- *Limitations: The paper has a limitation towards scope in terms of covering other aspects of logistics optimisation and various other factors impacting sustainabilty of Supply Chain.*

Keywords: Sustainable Supply Chain, Logistic Optimization, Green Supply Chain, Carbon Emission, Foot Print.

1. INTRODUCTION

Supply chain management business practice has been getting a growing attention both for the government and FMCG industries with regard to the environmental sustainability (*Sarkis, 2003*). There is a need to focus on integrating the strategic supply chain and operations plans with the environmental practices. Different successful initiatives have been taken by organizations to become environment friendly by inculcating the concepts of green supply chain with in the operations related to logistics or distribution or manufacturing or procurement (*Hasan, 2013*). The organization should employ an efficient supply chain in such a way that sustainability can be achieved right from procurement by screening the suppliers to monitor the environmental performance and do business with those who meet the regulatory standards (*Busch, 2013*). Introduction and implementation of different driving forces in the supply chain operations has been a proactive solution rather being a reactive approach to achieve environmental sustainability. Green supply chain management (GSCM) has an important role in addressing these environmental impacts that occurs in every stage of a product life cycle (PLC) (*Hutchison, 2003*). Thus if the enterprise implements GSCM then it will ensure that the company will have a surge in market share and profitability. This will be however possible by consciously evaluating the different environmental risks which exits in the overall supply chain. Like for the risk that is associated due to carbon accumulation in the food products during movement of finished goods (FG) or raw materials (RM), whether it is a perishable or non perishable packaged food and carbon emission in the environment due to the transportation of these food products.

A value chain network will be responsible for the development of storage and distribution of finished goods by the means of vehicle movement from factory to warehouses or hub and hub to warehouses depending upon the varied medium. Researchers have concluded that the a value chain network on which the freight movement and logistic operations executed faces five major challenges with respect to facility location model, distribution system for demand points, appropriate lot sizing for replenishment, bin packing optimal packing with in a container, vehicle routing for optimal FG allocation (*Friedrich et al., 2014*). Therefore it is necessary to design an efficient supply network which has a significant engineering importance in determining the facility location for manufacturing, storage and delivery of finished goods to the end consumers. This will result into a sustainable supply chains whose impacts in terms of carbon foot print and carbon emission can be measured and monitored by the industry (*Kauppinen et al., 2010*). It is important to design a perfect optimized distribution network by bringing the hub or spoke concept such that both producers and consumers take correct strategic decisions which will not affect the environment at any cost (*Wasner et al., 2004; Filho et al., 2013*). The reduction in freight transportation cost can directly underline recognizable relationship between the freight size type, medium of the delivery and number of kilometres travelled by the vehicle. Thus it is necessary to have a full truck load while transporting the FG from factory to the hub or warehouse or distribution centres (*Wieberneit et al., 2008*). The introduction of cross-docking stations makes the supply chain more environment friendly and eco friendly because the re-assembling and re-bundling constrain is removed which can reduce the carbon accumulation in the FG. Thus the overall delivery lead time reduces due to shipping and reduced intermediate logistic.

Since 1950, the advancement in logistic planning has been an important factor for sustainable supply chain due to the increase in globalization and nationalization of industries in various sectors (*Cooper et al., 1997*). So it is necessary that optimization of existing supply chain network will optimize the existing manufacturing and distribution processes depending upon the resources and management techniques that is available with the enterprises.

GSCM has apparently become an environmental innovation that integrates the environmental factors into complex supply chain management (*Zakuan et al., 2012*). Hence forth one of the key elements available in this complex supply chain management is distribution network which can be optimized to a larger extent my reducing the total freight cost in a transportation system. Freight cost is an integral part of transportation cost which implies one-third amount of total logistic cost. Thus the sustainability can be achieved by optimizing the freight cost which would become the prime factor to make the logistic chin green. The concept of GSCM discussed in this research paper, helps in reducing the number of kilometres travelled by the vehicle which will in turn, optimize the distribution network by reducing the overall freight cost. This ensures the reduction in carbon emission in the distribution chain which will in turn improve the operational excellence in the FMCG sector by proper delivery of finished goods to the consumer.This paper will basically focus on the concepts of Sustainable Supply Chain, Transportation and Logistic Optimization, Positioning of Warehouse and Hub Concept, Freight Cost Optimization. All these concepts have been combined strategically to acheive a suggestive model.

2. SUSTAINABLE SUPPLY CHAIN

Sustainable supply chain can be defined in terms of value chain which includes "sustainabili-ty" at all the steps from "farm to fork" (*www.dictionary.com*). Thus in other words the steps would include cultivation, processing, storage, transportation and distribution, consumption and waste disposal or recycle so as to maintain the relationship between the environment, economy and society. But, in this paper the concept of strategic supply chain has been restricted only towards Sustainable logistic chain.

With the number of food giant FMCG companies like ITC Ltd., HUL etc. are adopting sustainable value chain to create more long-term solutions for the planet which would be achieved by reduction of transportation cost of goods (*Sustainability Report HUL and ITC Ltd., 2014*). This in turn is providing an equitable employment and is improving the entrepreneurial spirit to design low-cost and effective solutions for Food Grain (FG) supply chain (*Hofstra, 2007*). The supply chain of the Indian companies are facing a lot of challenges in having appropriate efficient logistic systems to transport the goods, from farms to Mandi or production plants to process raw materials. Thus due to unorganized logistic operations and inefficient supply network whether transporting the RM or FG, the company incurs higher costs like freight cost which is very important factor to bring a sustainable supply chain. In contrast, it has been observed that, in a social entrepreneurial venture in the US or EU, the logistics needs are mostly for the delivery of the finished product and not for sourcing or procurement of raw materials (*ET Report, 2011*) (*Samarasinghe et al., 2013*). Henceforth this can be pros which can place a better sustainable value chain by effective consumption of the FG without any extra additional burden of transportation problems. The transportation problems can be classified in terms of the today's burning issues for company that is "FG Miles".

Five years post the outbreak of the world financial recession crisis, the global economy is still continuing to remain fragile and stagnate (*HUL Business Sustainability Report, 2013*). Thus to adopt a sustainable supply chain, company's firstly have to start in order to identify and evaluate climatic changes and demand fulfilment risks attached to different businesses. Secondly, reducing the impact of environmental parameters in the supply chain processes, product manufacturing and services; and thirdly, focussing towards creating a positive carbon footprint for major packaged food companies (*Felice et al., 2013*). This will improve in energy consumption, enhance usage of renewable energy sources and enlarge carbon positive footprint by expanding forest projects in wastelands (*Richard et al., 2013*). The FMCG companies are continuously working towards minimization of waste generation, maximization of reuse and recycling post-consumer waste as raw materials in the plan.

No to substantiate the sustainable supply chain concept GSCM is one of the efficient and effective sustainable supply chain solutions that the food companies are implementing in order to maintain the integration between both environment and business to customer (B2C) based supply chain management (*Kumar et al., 2013; Zenchanka et al., 2015*). It is also being identified and proved that sustainable supply chain leads to the innovation in new product specific technologies and improved supply chain performance. Companies are claiming that customer plays an important role in adopting sustainable supply chain in any sector (*ITC Sustainability Report, 2013*). A consumer demand directly pressurizes the company to produce a specialized packaged FG that in turn can be communicated to the primary vendor to procure to supply the appropriate RM. It clearly indicates that the consumer is indirectly using as the natural resources. Thus, we can conclude that consumer requirement and need has more value proposition in making the supply chain as sustainable by exploiting the environmental opportunities.

Whenever the question arises with respect to managing supply chain with respect to sustainable operations then the concept of, Third Party Logistics (3PLs) can help consumers, reduce costs involved during shipping of materials by providing proper guidance on package operations and design (*Green Biz Report, 2014*). Let us consider a food industry whose major objective is to identify the most efficient type of packaging design and materials which can be used to package the FG. This packing material should have the ability to be recycled or reused and does not harm the nutritional content of the food product while movement of the FG during cross docking. To manage a sustainable supply chain efficiently, it is mandatory that the positioning of warehouses and distribution centres should be decided by keeping the sustainable factors in mind (*Perl et al.,* 2009).

3. TRANSPORTATION AND LOGISTIC OPTIMIZATION

The solution lies in building a sustainable transportation or logistic system which will incorporate the 3PL concept that provides an efficient, sustainable supply chain solution such that the end customers are able to mitigate environmental impact in a number of ways beyond key performance measurement (*Vermeulen, 2010*). With the availability of a broad range of sustainable technologies as well as transportation and warehousing optimization solutions, certain 3PLs can assist the company and its primary customers to implement sustainable supply chain strategies such that heavy cost savings can be attained parallel to growth of company (*Azevedo et al., 2011*). This approach will include to access and use the available transportation and vehicle scheduling options which are less in carbon footprint-intensive such as rail, truck or ocean carrier modes or having a partnership with transportation service providers or carriers by investing to design and manufacture or release a particular eco-friendly hybrid quality or natural gas vehicles within the geography (*Wong, 2012*).

From the technology perspective, 3PLs can deal their customers (food companies) to incorporate and implement paperless solutions required for commercial invoices like purchase orders, goods invoice, goods receipt, delivery Challan and billing, as well as customs and export order documents, to reduce overall paper consumption in order to have smooth business operations. One of the important areas where a 3PL solution can make an extraordinary huge impact on having sustainable supply chain is by allowing customers to set up an efficient reverse logistic network operation *i.e.* from customer to source. It is necessary for food companies to implement different sustainable processes for recycling and refurbishing the product or product waste or proper disposal techniques during product exit in a product life cycle which has led to a significant reduction in acquiring total carbon footprint (*Lee, 2012*). This can be more clearly understood by considering today's distribution challenges, whether for perishable or non perishable food product, 3PLs is one of the only sustainable solution which can offer the effective services by centralizing the warehouse or distribution centres that will make a huge impact to an extent by reducing the number of miles travelled to reach the final consumer in a food supply chain (*Selviaridis, 2007*). However, after being implemented these solutions, the company will be still held responsible for accumulation of certain levels of carbon impact in a value chain. Currently Indian FMCG sectors are more focused towards investing in environmental projects that includes goods indent carriers, which has a direct impact on shipments (*Cardoso et al., 2015*).

3.1 Case Study – TTKL Logistic Optimization

One of the case studies illustrates that how 3PLs can impact to bring sustainable supply chain. TTK logistics (TTKL) is a 3PLs company which was established in December 2002 in order to manage the Milk Run supply chain of Toyota Motor Thailand (TMT) in Bangkok (*Nemoto et al., 2013*). Milk Run logistics is a scenario when there is a demand for services that is frequently repeated with respect to just-in-time or just-in-sequence (JIT/JIS) at multiple pickup or delivery points and when the company or service provider is able to customize the vehicle size as per the demand. The TMT company operation in Milk Run logistics comprises of locally procuring different automobile parts from different positions and includes an optimal route planning. In other hand logistics operation comprises of only stock transfer within the value chain. Currently, the Milk Run logistics model is implemented in four factories of TMT with around 50 distinct delivery routes established for each plant. Since Milk Run logistics impacts the automobile production planning, thus a close cooperative relationship is established between suppliers, TMT and TTKL. TMT shares daily production and operational plan, automobile parts pickup and delivery point information to TTKL such that perfect order fulfilment and on time delivery and full are achieved.

Efficient utilization of Milk Run logistics increases the average loading rate that results in less environmental carbon emission. High loading factor can be realized by TMT company because the truck indent was full every time (supplier sends the automobile parts to TMT and the TMT uploads the defective parts back to the supplier post unloading by same truck). Based on detailed logistic operational data, TMT estimated that carbon emission due to Milk Run logistics compared to the situation without Milk Run logistics was reduced up to 13.6 tons carbon dioxide emissions per day and per factory, which is about 53% reduction (*Hashimoto, 2009*). Thus this indicates that by just implementing 3PL solution the overall logistic operation is optimized by having a proper optimized supply network in order to make supply chain green and sustainable.

(Extracted from Milk Run Logistics in Bangkok; Nemoto *et al.,* 2013)

4. POSITIONING OF WAREHOUSE AND HUB USING LOGISTIC OPTIMIZATION

To understand a hub-and-spoke logistics network, it is necessary to understand the primary functions of a hub whose prime objective is to perform transhipment operations. This means the re-directing and re-assembling of big consignments into smaller consignment. Spokes or distribution centres or warehouse are the end node in a logistic chain which links the end customer with the hubs (Zapfel *et al.,* 2002). Usually, the distribution processes in a hub and spoke network model can be summarized below:

Stage 1: Factory depot that is located in a supply chain network will pick up the shipments for their respective customers. This can be done by carrying the shipments for several customers called as pick up tours in the same round trip within a particular region. The decision for number of pick-up tours or the routing of this pick up tour will depend upon the operational decisions and network optimization. At the end of this stage it is mandatory that all the ordered shipments are being available at the factory depot for the movement to the hub.

Stage 2: This is one of the complex stage which comprises of:
- Transportation of FG from factory depot to the hub;
- Cross-docking, re-arrangement and transhipment of FG from several factory depots to assemble into a new consignments which has the same destinations as others have;
- Transportation of FG from hub to the warehouses or the destination depots.

Usually to keep make sure that the vehicle moves less number of miles, it is necessary that the same vehicle carrying the consignment from factory depot to hub and hub to destination depot should happen subsequently. This is ensuring that the milk run distribution process is followed. The logistic operations excellence depends upon managing and balancing the outbound and inbound logistics traffics simultaneously (*Ganesh, 2012*). Thus the constraint delivery time can be reduced by adopting the milk run supply chain by simply temporary load balancing (holding certain consignments).

To perform these arrangements, it is mandatory to have sufficient haulage capacity which is only possible with the number of vehicles available in that particular route. This is one of the factors which makes sure that as per the availability of trucks the hub and warehouses are positioned for a region in order to fulfil the demand. The positioning if the hub and warehouse in a supply chain execution network also depends upon the requirement of extra vehicles during factory clearance. This concept makes us understand that the distribution network should be designed in such a way that the number of kilometres travelled by the vehicle from factory depot to hub to warehouse should be lesser. Thus the emission of carbon to the environment will be lesser extent (*Saikku, 2009*). This adds a positive carbon foot print keeping the supply chain environmental friendly and sustainable.

Stage 3: Once the vehicle reaches the destination depot or warehouses from the hub then the same stage 1 step is followed to make sure the maximum utilization of the truck by delivering the FG at the destination hub and then have the pickup route in case of returning the FG to hub or moving some FG to any other warehouse which falls in the same route.

This whole hub and spoke model dignifies the route optimisation by right vehicle scheduling and reducing the delivery lead time. This distribution network will not only improve the profitability in terms of freight cost but also will help to maintain the sustainability of the logistic chain in a complex distribution network.

To ensure that the supply chain is sustainable, the prime impact parameters required for logistic optimization deals with increased focus on freight cost perspective along with services required during planning and design of value chain network (*Camargo, 2009*). The initiative goes in order to find the most cost optimized way in meeting key objectives for an excellent and effective supply chain.

To make a lean supply chain, the major challenge lies in optimizing the value chain that entails by relooking at the fundamentals of business model (*Sharma, 2013*). This ideally impacts many areas of supply chain of the enterprise like local supplies vs. imports; Interstate or intrastate procurement of raw material or semi FG and services; production, processing and location of warehouse; own or contract based manufacturing; Direct sale vs. stock transfers.

Since we know that warehouse is an important and integral part of any supply chain. Thus placing the warehouse strategically will not only improve customer service levels but also will reduce the number of miles travelled by the vehicle in a value chain.

5. FREIGHT COST OPTIMIZATION MODEL

Value chain Network Design is a process to determine a unique network configuration for a sustainable supply chain which offers the lowest total cost/ highest total profit, considering operational risks and financial risks while achieving targeted Service Levels. To illustrate the above explained model of hub and spoke concept, milk run supply chain and freight cost optimization in significant to sustainable supply chain, we propose a model that clearly indicates that using 3PL services the logistic network optimization can be done.

Let us assume that Guntur is the turmeric plant which is supplying the FG to destination depots or warehouses in Siliguri, Vizag, Vijaywada, Chennai, Hyderabad, and Hubli with a freight cost of ₹ 9.10, ₹ 2.80, ₹ 1.70, ₹ 3.0, ₹ 1.10 and ₹ 4.40 respectively per carton (bin packing) of FG as per the following network design in Figure 1.

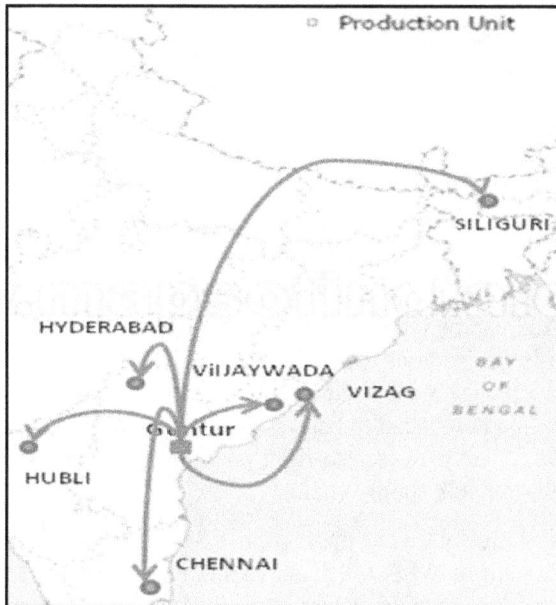

Fig. 1: Actual Distribution Network

We can easily observe that the freight cost for sending the FG from mother plant or the factory depot Guntur to Siliguri and Hubli is highest per carton. Thus the total transportation cost calculated for these two regions for a particular indent of truck will be higher than any other region because of the higher freight cost. Now the COGS (Cost of Goods Sold) will be calculated based on the Cost of goods manufactured (COGM) and the freight cost associated with it to transfer the goods from factory depot to the warehouse service provider (WSP) or distribution centres (DC). As per the freight rate described in Table 1, because of the higher freight cost from Guntur to Siliguri and Hubli, the margin for the distributor falling in these regions will be lesser. This in turn will lead to higher MRP (Maximum Retail Price) for end customers. Higher freight cost is not only the consequences of increased selling price but also repeated freight movement within the state and beyond the state can lead to increase in the price of FG. This is because of the fact that the repeated movement of the vehicle in the above supply network will incur extra local and Octroi taxes in addition to the COGS. Thus the existing network is not at all sustainable as per company's vision and even carbon emission is higher due to the larger distance movement without any cross docking.

Let us assume that the demand for the destination depots as per the network from Guntur factory will be as follows in Table 1.

Table 1: Freight Cost and Estimated Demand for the Month of April 2015

Factory Depot	WSP/DC	Demand for April 2015 (in Cartons)	Freight Cost (Rate/Carton)	COGM (in ₹)
Guntur	Siliguri	242	9.1	2203
Guntur	Vizag	16961	2.8	47660
Guntur	Vijaywada	17742	1.7	30161
Guntur	Chennai	1359	3.0	4008
Guntur	Hyderabad	689	1.1	758
Guntur	Hubli	676	4.4	2973
Total Freight Cost		**37668**		**87763**

Now a cost saving distribution model can be proposed to the existing network such that the total freight cost ₹ 87,763 can be reduced. In case of fluctuating demand, it is necessary that a lot size inventory should be made at the WSP and DC for immediate replenishment for the FG. To have a proper replenishment cycle of FG, the availability of truck for the particular zone, reduced freight cost and reduced transit inventory have to be focussed to a greater extent (*Wieberneit et al., 2008*).. This will lead to right positioning of hub or distribution depot in a supply network and a sustainable supply chain.

Major FMCG companies are following the Hub and Spoke model for the distribution of FG. Hub is a kind of storage unit or WSP which act as a company owned depot to store the FG to deliver the customer demand at right delivery time. Hub acts as the secondary plant for a company which holds a lot size inventory and is responsible to replenish the order periodically received from the distribution depots or ware house.

Thus our main objective is to position a hub somewhere near Kolkata and Bengaluru where the cross docking of the shipments can be done periodically depending upon the demand. These hubs can act as the mother plant for both Siliguri and Hubli respectively. The optimized sustainable network can reduce the total freight cost and the overheads incurred during the transportation will be lesser. The hub Kolkata will receive the shipment from the factory depot present in Guntur and can even service Vizag and Vijaywada if the demand is not replenished directly by the factory depot Guntur.

Thus a milk run supply chain can be executed in the optimized network where the truck with full load can collect all the shipping consignment for hub Kolkata, warehouses Vizag, Vijaywada and Siliguri from the Guntur depot and follow the following subsequent steps:

- Movement of FG from Guntur depot to Kolkata hub by disposing the FG at Vizag and Vijaywada (for direct shipments to these warehouses which falls under the shipping route)

- Cross docking at Kolkata hub and movement of full truck load either for Siliguri or a full truckload for Vizag and Vijaywada.

We can analyze the same by identifying the total freight cost from Guntur factory depot to Kolkata Hub or Vizag and Vijaywada and from Kolkata to Vijaywada or Vizag or Siliguri as per the optimized network with hub and spoke concept shown in Fig 2. It may also happen that the freight cost per carton from Kolkata hub to Vizag, Vijaywada, Siliguri will be lesser or more than the freight cost per carton from Guntur depot to respective warehouses. But the total freight cost incurred in the optimized network will be lesser than the non optimized network. Any vehicle which is to be indented for a hub should always have a full truck load and should follow a milk run supply chain distribution model. The reason behind having this optimized model will certainly reduce the total transportation cost and number of miles travelled by the truck. This will finally impact the overall sustainability of supply chain by increasing the profitability and reducing lead time due to logistic operations.

Another important factor for positioning the hub at Kolkata is because of the higher availability of the truck to destination depots in this zone compared to Guntur to Vijaywada and Vizag and Siliguri.

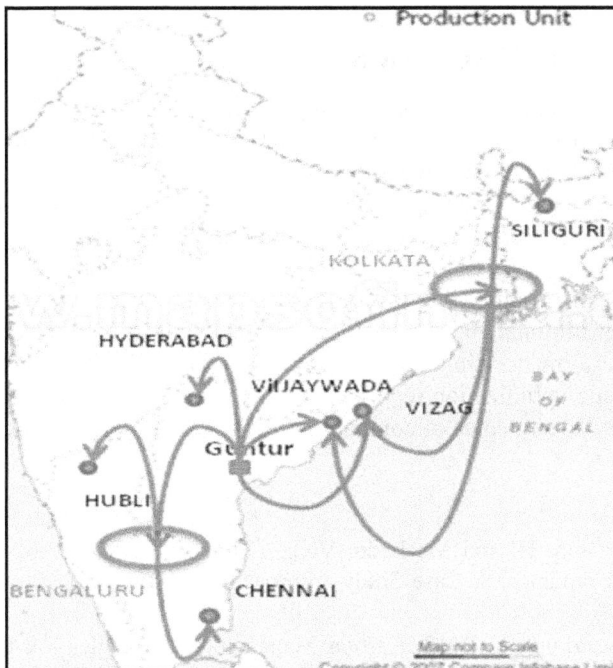

Fig. 2: Optimized Sustainable Distribution Network

The similar supply network execution is done by positioning the hub in Bengaluru to fulfil the demand of Chennai and Hubli destination warehouses. The total freight cost calculated on the basis of the new optimized network is shown in Table 2 which is ₹ 1,858 less than the actual network.

Table 2: Total Freight Cost for Optimized Network

Factory Depot	Hub	WSP/DC	Estimate for April 2015 (in Cartons)	COGM (in ₹)
Guntur	Kolkata	Siliguri	242	705
Guntur	Kolkata	Vizag	16961	49186
Guntur	Kolkata	Vijaywada	17742	31758
Guntur	Bengaluru	Chennai	1359	2228
Guntur	Guntur	Hyderabad	689	758
Guntur	Bengaluru	Hubli	676	1270
Total Freight Cost			**37668**	**85905**

From the above Table 2, this may be ensured that the logistic optimization can be achieved by changing the supply distribution network of the enterprise such that the total freight cost reduced. This not only reduces the overall transportation cost but also reduces the number of kilometers or miles travelled by the vehicle. Thus the emission of carbon to the environment reduces and the sustainability is maintained through out the supply chain in terms.

6. SUGGESTION AND CONCLUSION

The suggested model supports sustainable supply chain in terms of maximized customer service, minimum transit cost and lesser time in transit (TIT) to full fill the unconstrained demand. However, this cannot be generalized that logistic optimization is the only solution which can be proposed in 3PL arrangements to make the supply chain sustainable. There are many other factors which can improve the supply chain in terms of carbon emission and carbon foot print. Logistic optimization illustrates that how supply chain network design exercise helps in finalizing the Zone of indifference/Range of indifference where the supply chain cost becomes minimal on the basis of various tradeoffs involved. During network design exercise a performance simulation can also be adopted by the industries by generating the sustainable logistic optimization reports that talks about the carbon emission and validates the feasibility of GSCM strategies of company for acheiving environmental sustainability.

REFERENCES

Azevedo, S.G., Carvalho, H. and Machado, V.C., 2011, The Influence of Green Practices on Supply Chain Performance: A Case Study Approach, *Transportation Research Part E* 47(6): pp. 850–871.

Busch, T. and Schwarzkopf, J., 2013, Carbon Management Strategies—A Quest for Corporate Competitiveness, *Progress in Industrial Ecology, An International Journal*, Vol. 8(1/2): pp. 4–29.

Camargo, R., Miranda Jr., G., Ferreira, R. and Luna, H., 2009, Multiple Allocation Hub and Spoke Network Design Under Hub Congestion, *Computers and Operations Research* 36: pp. 3097–3106.

Cardoso, A.V., *et al.*, 2015, Carbon Market and Global Climate Governance: Limitations and Challenges, *International Journal of Innovation and Sustainable Development*, Vol. 9(1): pp. 28–47.

Cooper, M.C., Lambert, D.M. and Pagh, J.D., 1997, Supply Chain Management: More than a New Name for Logistics, *International Journal of Logistics Management*, Vol. 8(1): pp. 1–13.

Felice, D.F., *et al.*, 2013, An Integrated Conceptual Model to Promote Green Policies, *International Journal of Innovation and Sustainable Development*, Vol.7(4): pp. 333–355.

Filho, W.L., 2013, Applying An Ethical Decision-Making Model: The Case of Avco Environmental, *Progress in Industrial Ecology, An International Journal*, Vol. 8(3): pp. 135–144.

Friedrich, H., and Gumpp, J., 2014, Simplified Modeling and Solving of Logistics Optimization Problems, *International Journal of Transportation*, Vol. 2(1): pp. 33–52.

Ganesh, K., *et al.*, 2012, Study and Suggestions on Inbound Logistics for Vehicle Utilisation: A Case Study, *International Journal of Innovation and Sustainable Development*, Vol. 6(3): pp. 237–264.

Hasan, M., 2013, Sustainable Supply Chain Management Practices and Operational Performance, *American Journal of Industrial and Business Management* 3: pp. 42–48.

Hashimoto, M., Ishihara S., Nemoto, T. and Inaba, J., 2009, Logistics Management of Automotive Parts in South China, *Journal of Japan Logistics Society*, Vol. 17: pp. 161–168.

Hofstra, N., 2007, Sustainable Entrepreneurship in Dialogue, *Progress in Industrial Ecology, An International Journal*, Vol. 4(6): pp. 495–514.

Hutchison, J., 2003, Integrating Environmental Criteria into Purchasing Decision: Value Added, In: T. Russel, Ed., Green Purchasing: Opportunities and Innovations, *Green-leaf Publishing, Sheffield*, pp. 164–178.

Kauppinen, T., *et al.*, 2010, Carbon Footprint of Food-Related Activities in Finnish Households, *Progress in Industrial Ecology, An International Journal*, Vol. 7(3): pp. 257–267.

Kumar, S., Luthra, S. and Haleem, A., 2013, Customer Involvement In Greening the Supply Chain: An Interpretive Structural Modeling Methodology, *Journal of Industrial Engineering International* 9(6): pp. 1–13.

Lee, K.H., and Kim, J.W., 2012, Green New Product Development and Supplier Involvement: Strategic Partnership for Green Innovation, *International Journal of Innovation and Sustainable Development*, Vol. 6(3): pp. 290–304.

Nemoto, T., 2013, Efficient And Green Logistics in Urban Areas—A Case of Milk Run Logistics in the Automotive Industry, *Hitotsubashi University, Tokyo, Japan*, pp. 1–16.

Perl, E., and Vorbach, S., 2009, Environmental Information for Sustainable Supply Chains, *Progress in Industrial Ecology, An International Journal*, Vol. 6(1): pp. 44–67.

Richard, R., Rushforth, Elizabeth, A., Adamsa and Benjamin, L., Ruddell, 2013, Generalizing Ecological, Water and Carbon Footprint Methods and Their World View Assumptions Using Embedded Resource Accounting, *Water Resources and Industry* Vol. 1(2): pp. 77–90.

Saikku, L., 2009, Consumers and Macro-Level Forces Behind CO_2 Emission Development, *Progress in Industrial Ecology, An International Journal*, Vol. 6(4): pp. 371–386.

Samarasinghe, G.D. and Samarasingh, D.S.R., 2013, Green Decisions: Consumers' Environmental Beliefs and Green Purchasing Behaviour in Sri Lankan Context, *International Journal of Innovation and Sustainable Development*, Vol.7(2): pp. 172–184.

Sarkis, J., 2003, A Strategic Decision Framework for Green Supply Chain Management, *Journal of Cleaner Production*, Vol. 11(4): pp. 397–409.

Selviaridis, K. and Spring, M., 2007, Third Party Logistics: A Literature Review and Research Agenda, The International Journal of Logistics Management, Vol. 18(1): pp. 125–150.

Sharma, K., 2013, A Case Study On Mcdonald's Supply Chain in India, *Asia Pacific Journal of Marketing and Management Review*, Vol. 2(1): pp. 112–120.

Vermeulen, W.J.V., 2010, Sustainable Supply Chain Governance Systems: Conditions for Effective Market Based Governance in Global Trade, *Progress in Industrial Ecology, An International Journal*, Vol. 7(2): pp. 138–162.

Wasner, M. and Zapfel, G., 2004, An Integrated Multi-Depot Hub-Location Vehicle Routing Model for Network Planning of Parcel Service, *International Journal of Production Economics* 90: pp. 403–419.

Wieberneit, N., 2008, Service Network Design for Freight Transportation: A Review, *OR Spectrum* 30(1): pp. 77–112.

Wong, D.W.C., *et al.*, 2012, An Intelligent Vehicle Management System for Reducing Greenhouse Gas Emission—A Case Study of Hybrid Vehicle Engine, *International Journal of Innovation and Sustainable Development*, Vol. 6(4): pp. 347–367.

Zakuan, N., *et al.*, 2012, Green Supply Chain Management: A Review And Research Direction, *International Journal of Managing Value and Supply Chains (IJMVSC)*, Vol. 3(1): pp. 1–18.

Zapfel, G. and Wasner, M., 2002, Planning and Optimization of Hub-And-Spoke Transportation Networks of Cooperative Third-Party Logistics Providers, *International Journal of Production Economics* 78: pp. 207–220.

Zenchanka, S. and Korshuk, E., 2015, The Green Economy Concept in Belarus: Today and Tomorrow, *Progress in Industrial Ecology, An International Journal*, Vol. 9(1): pp. 33–45.

Analysis of the Corporate Social Responsibility Initiatives in the ICT Industry Using Triple Bottom Line Approach

Sujata Joshi[1], Gaurav Sethia[2], Abhishek Sharma[3],
Abhinav Nirwan[4] and Jaspreet Singh[5]

Symbiosis Institute of Telecom Management, Symbiosis International University, Pune
E-mail: [1]sjoshi@sitm.ac.in; [2]gksethia@gmail.com; [3]abhishek.sharma@sitm.ac.in;
[4]abhinavnirwan2012@gmail.com; [5]jaspreet.singh@sitm.ac.in

ABSTRACT: *Corporate Social Responsibility (CSR), in recent years, has gained immense importance, as organisations have understood the fact that what is beneficial to the society is beneficial to the organisation. It has been observed that organizations which actively promote their corporate social responsibility initiatives are normally viewed more favourably than organizations who do not have such initiatives After the compulsory inclusion of corporate social responsibility in the Companies Act, 2013, organisations have become more active and have started investing in this area. CSR initiatives help companies gain positive social media visibility, better customer engagement and also build a positive work place environment.*

Aim: *This paper analyses some of the CSR initiatives taken up by Information, Communication and Technology (ICT) companies in India and its impact on social, environment and economic performance of the company using triple bottom line framework.*

Methodology: *The paper uses case study approach for analyses of the CSR activities of ICT companies. The companies selected here are rated as top four companies as per CSR rating given by the research done by IIM Udaipur for top companies in CSR in different domains.*

Findings: *The research in this paper states the initiatives taken by ICT companies and its impact on the triple bottom line of the company's social, environmental and economic parameters.*

Managerial Implications: *The paper will be useful for ICT managers to realize the benefits of the CSR activities identify solutions to the challenges and come up with innovative strategies to make their business sustainable.*

Originality: *Although CSR initiatives have been undertaken by companies, very few studies have been conducted especially in the ICT area on analysis of CSR initiatives. This paper attempts to fill this gap.*

Limitations: *The paper focuses on only ICT companies and the sample size is also a constraint as the study considers company information from secondary sources.*

Keywords: Corporate Social Responsibility (CSR), Triple Bottom Line, (ICT) Information Communication and Technology, Sustainability.

1. INTRODUCTION

The field of Corporate Social Responsibility has grown a lot in the last decade and after the compulsory inclusion of CSR in the Company's Act 2013 the companies have started investing in these activities positively. Companies have started investing in different initiatives in the fields like social, environment, health and educational development. Now a large number of companies than ever before have started making provision for corporate social responsibility in their business model and are putting sincere efforts in doing so.

The definition of Corporate Social Responsibility has been difficult to understand. As per (McWilliams and Siegel, 2000) "actions or steps taken in the direction of social betterment and which are beyond the benefit of organisation." Some state that "there is a strong relationship between the initiatives taken by company as CSR and consumers behaviour towards it." (Bhattacharya and Sen, 2003). "The initiatives intended for the good of others with no immediate reward for the benefactor" (Organ, 1988). In simple words CSR can be defined as the responsible initiatives take by the companies towards the social and economic development keeping in mind the betterment of environment.

1.1 Benefits of CSR

A socially responsible company has a better brand image and reputation the consumers usually tend towards companies with better reputation and image (Margarita, 2004). A company which is considered as socially responsible also has its reputation in the business community and has a higher chance to attract capital and trading partners. There are studies which indicate that CSR initiative positively influence and drive the consumer decisions (Kim and Park, 2009). CSR is a way to create brand equity of the organisation as it adds value to the organisation which directly improves its financial performance (Mullen, 1997). The recent research states that there is a direct relationship in company's CSR initiatives and the attitude of customers towards the company (Brown *et al.,* 1997); (Creyer *et al.,* 1997); (Ellen *et al.,* 2000). Hence CSR initiatives develop a positive image of the company in the consumers mind.

2. CONCEPTUAL FRAMEWORK

The Triple Bottom Line (TBL) approach is used as a framework for measuring the performance of the firm against its social, environment and economic performance. This paper works on the attempt to align the ICT industry with the Triple Bottom Line approach and measure its CSR initiatives and their impact on the sustainable development of the company. It is one of the initiatives to provide the industry with more comprehensive objectives than just earning profits. The perspective taken here is that for any organisation to have a sustainable growth it should be financially sustainable, it should have minimum negative environmental impact and also should be in line with the societal expectations.

3. RESEARCH METHODOLOGY

In order to analyse the Corporate Social Responsibility initiatives of ICT companies, the researcher has used the case study approach. The researcher has selected four ICT companies namely Infosys Limited, Wipro Technologies, Bharti Airtel and Tata Consultancy ltd. These four companies have been selected for this study as they are rated as top four companies as per CSR rating given by the research done by IIM Udaipur (2014), for top companies in CSR in different domains.

The below mentioned sections deal with the analysis of the Corporate social responsibility companies of the four companies and its impact on social, environment and economic performance of the company using triple bottom line approach.

4. ANALYSIS OF THE CSR INITIATIVES

4.1 Infosys Limited

Infosys envisions a vigorous tomorrow which has a healthy society, rich ecosystem and greener industrial and domestic practices. To make it a reality the company has a sustainable policy for the society which comprises of various stakeholders like local communities, social organizations, clients, government and non-governmental authorities. The sustainable policy is not only with respect to entities but it also focuses on resource optimization, operations, product, supply chain, processes, service models, stakeholder engagement and branding. The policy focuses on four main aspects:

1. Business sustainability
2. Client's business sustainability
3. Ecosystem sustainability
4. Lifestyle sustainability.

So according to the triple bottom line framework approach of Corporate Social Responsibility, there are three main pillars/parameters to measure the impact of the CSR initiatives on the company performance: these are Social, Environmental and Economical parameters. We shall discuss each one of them with respect to the CSR initiatives taken by Infosys and what has been the impact of such initiatives.

4.1.1 *CSR Initiatives of Infosys Limited*

Area	CSR Initiatives	Impact
Social	• SPARK • Influence • Parishudh • Infosys Foundation • HALE Health weak	• 6,20,000 households covered for the rural development program • Trained 30% of suppliers on Supply chain management • Raised aspirations of over 159827 students through SPARK program • Enabled 1692 faculty members and 60809 students of campus connect

Environmental	• Reducing ecological footprint • Digitization of assets	• Reduction in carbon by 53.53% as compared in 2008 • 40% reduction in carbon intensity • 34% reduction in water intensity • Highest Green building certifications covering 2 million sq.
Economical		• Brand valuation increased to 12,596 from 11063 • Revenues increase from 6994 to 7398 • Number of clients increased to 798 from 694

4.1.2 *Social, Environmental and Economic Impact of CSR Initiatives*

Social

Social section of the framework deals with industry-academic partnership to nurture fresh talent, employee engagement programs to provide growth opportunities to existing employees, increasing client value and community empathy etc.

The company has been nurturing talent through its Talent Enablement program which covers foundation program, continuous education, collaborative learning, research, higher education and credit points for all the employee categories like Associate, Middle, Senior, Top management. It creates an ecosystem for knowledge management which covers major aspects like advanced technology, new processes, and new business initiatives to keep up the competencies of the employees. The company has a Competency Development Program as well which offers relevant opportunities and flexi role along multiple competencies. Internal career mobility policy has been re-launched in 2013 to facilitate promotions and progressions.

Various initiatives like collaborative learning programs like 'KM Portal' a knowledge repository and 'Konnect' a professional networking platform used by 1,00,000 employees, Team wiki, KMail, *etc.* help in knowledge management of employees. Higher education is being provided in association with U.K University to the 232 new professionals making it a total of 932 employees at the end of year 2012–13.

SPARK program launched in 2008 has 4 main programs in its portfolio: On-campus events, rural reach program, catch them young, GURU. These programs have been organized 3206 events, trained 5,41,766 students out of which 2,05,986 are girls and 1,78,187 from rural areas and under the GURU program a total of 27,004 faculties have been trained at the end of 2012–13.

In 2013 the company has also set up Influence i.e. Infosys Framework for learning using external community engagement which helps in recognizing and cherishing volunteering spirit and community development at Infosys. This initiative let the employees work for the society in their free time which gives them additional credit points under Competency Development Program 3.0. Since it's a new initiative the impact is yet to be measured.

Infosys annually contributes 1% of its PAT to Infosys Foundation which supports underprivileged section of the society. Foundation has helped in building 40,000 school libraries by contributing US $ 1033,172 in Karnataka. Also the foundation has invested US $ 1,833,517 and started Parishudh initiative which has helped 10,000 families to improve their sanitation problems.

Environmental

A significant reduction in the consumption of resources is one of the main objectives of the CSR initiative of the company. The company has made efforts to reduce its electricity consumption by 40% and water consumption by 34% as compared to 2008. The company plans to source its electricity consumption from renewable resources by 2017 which is 22% in the portfolio according to reports of 2012–13.

The company also believes in building smart buildings/offices because they believe that buildings are the biggest energy consumers and 33% of the CO_2 emission comes from buildings.

It starts with a small step and steps like they replaced 250W sodium vapour lamps with 75W LEDs resulting in 70% reduction on the overall load. Installing Daylight sensors and occupancy sensors in the office helps in reducing operational cost. Efficient Air-conditioning technology helps in reduction of 5–8% efficiency.

They have also contributed to the greenery and biodiversity conservation as well by planting 62,065 saplings and there are about 2,88,065 trees in all the campuses of Infosys.

Following initiatives has led the company to achieve 53.53% reduction in GHG emissions. The strategic partnerships with various other organizations like united technologies Research Centre (UTRC), Saint Gobain Research India Limited and Indo-US joint centre for building energy research and development.

Economical

Though no specific initiative was taken to improve the financial performance of the company but the researchers believe that the Social initiatives taken by the company has indirectly affected the financial performance of the company.

The above statement can be supported by the juncture of performance like Brand Valuation, Economic value added, Net profits, etc.

The snapshot of the financial performance of the company shows that out of the total revenues and other income the company has contributed 52.85% in the employee wages and benefits. Also, The Company has contributed US $ 2 million in the Infosys Foundation which was used for Social and Environmental causes as stated in the above paragraphs.

The contribution to the society has resulted in increase in Brand valuation of the company from US $ 11,063 million to US $ 12,596 million. The brand valuation shows how the Brand Infosys is viewed in the market. Also, the total revenue of the company has increased from US $ 6,994 million to US $ 7,398 million in 2013. The net income has increased from US $ 1,716 to US $ 1,725 million. We believe the contributing factor to this growth in Brand Valuation, Revenues and Net Income is the fact that the company has reduced its operational cost by focusing on renewable energy simultaneously. As it has been adopting the energy efficient technology its operational cost has decreased.

Moreover, the focus on the relationship maintenance with its suppliers has led to increase in number of clients from 694 to 798 in the year 2012–13. It is because the company has improved the supplier management cost by giving free consultation to its main suppliers.

4.2 WIPRO Technologies

Wipro's approach towards sustainability has focussed on enabling itself as an organisation and its customers to be more ecologically sustainable. It has always considered issues which are important to employees, customers, suppliers, investors etc. According to Wipro, "Now they possess convincing contents to leverage social issues which are the "power to do well". Purpose of Wipro CSR guidelines:

1. Answer latest issues of current generation
2. Helping in building a good society for coming generations.

4.2.1 *CSR Initiatives*

Area	CSR Initiatives	Impact
Social	• WIPRO Cares • Applying Thought in Schools • Mission 10X • Siyapha	• Access to more than 75000 people to primary health care • Worked with around 2000 schools going up to 800,000 students • Covered around 1200 engineering colleges over 25 states in India • Completed 4 batches of graduate training programmes for the client, covering over 200 South African employees
Environmental	• Eco Eye • Earthian • Participative Community Water Program • Rapid Development of Renewable Energy (RE) solutions	• Admittance in DJSI and NASDAQ100 • Around 3000 schools and colleges enrolled in the program • Developing a detailed groundwater aquifer map for a 33 sq. km area around our Sarjapur • Procured 66 Million units of RE translating into 52000 metric tons of greenhouse gas emissions avoided
Economical		• PAT increased from 52,325 Mn to 61,362 MN in the financial year 2012–13 • Voluntary Attrition reduced to 13.7% from 17.5% • DPS increased from ₹ 6 to ₹ 7 in the financial year 2012–13

4.2.2 *Social, Environment and Economic Impact of the CSR Initiatives*

Social

WIPRO under the initiative Wipro Cares emphasized on communities which were deprived of basic facilities. Started in 2003, presently working for 16 projects. With its various healthcare projects in various states of India Wipro Cares is contributing to more than 75000 people in 53 villages by providing them primary health care. With various projects in disaster rehabilitation, the have helped in revitalising lives of many people injured in floods. With its education projects it helped more than 47,000 children in their education. "Applying School of thoughts" initiative, the motive was to advance learning in schools. The Mission 10X worked towards bringing change through partnerships with civil society organisations. This initiative introduced Education literature program and Advocacy program for the betterment of graduates in India with the participation of scholastic and educational institute in India. Mission 10x has covered 1200 engineering colleges over 25 states in India.

Environment

CSR initiative named Eco Eye was undertaken by Wipro, with an objective to strike a balance with the environment, towards sustainable business practices. The operation mainly focussed on energy and water productivity. Other areas were interconnection of sewer line, rainwater harvesting and managing solid waste by setting bio-gas plants.

Earthian, a sustainability program for schools and colleges, focusses on engagement of students towards environmental issues. 3000 schools and colleges have been registered in the program so far. Every time they give new topics and themes to students to discuss. Schools were given water as a theme while colleges were given global warming as an issue. Participative Community Water Program, initiated in the Sarjapur area in Bangalore which is completely dependent on groundwater, seeks to involve in a community led governance model of groundwater. In the first phase a detailed groundwater aquifer map for a 33 sq. km area around our Sarjapur campus was successfully developed.

Economic

The electricity consumption, CO_2 consumption and water consumption has reduced by 13.1%, 9.4%, 13.8% respectively proves that the company has reduced its operational cost by using renewable source of energy and decreasing its dependency on the non-renewable factors.

Also the management approach of improving the supply chain by proliferating industrialization in manufacturing and disintegration of services through IT services has helped in acquiring economies of scale and hence helps in reduction of operating cost and it offers the company higher profit margins. This improves the financial statements of the company.

Wipro is the only company in the top 4 section to have a negative goodwill. It represents the negative relation between the CSR initiatives and financial performance of the company.

The Earning per share (EPS) of the company increased from ₹ 19.08 to ₹ 22.99 in the financial year 2012–13 according to the financial report again signifying the surge in profits for the company and hence higher EPS to the shareholders but the decrease in valuation of the goodwill still bewilders the TBL equation.

4.3 Bharti Airtel

Airtel presents its "Outline for Social Inclusion" in the 2012–13 report, which was created after a venture of a few "happy and lively worker hours". The outline incorporates three columns, each of which consisting of a vision and an action approach. A supplier eco-framework chart helps comprehension of the complex interfaces and associations needs to keep up manageable operations, and the worker engagement area is a reviving take a gander at how this organization backings and enables representatives.

The Annual Report of the Bharti Foundation is the only common disclosure mechanism of the Corporate Social and Environmental Reporting by the Bharti Group. This is the only process of evaluation of the CSR Agenda. There has been no third party involvement for environmental and social auditing. Unlike others, Bharti Group as such does not follow any GRI or similar guidelines for reporting standards.

4.3.1 *CSR Initiatives*

	Initiative	*Outcomes*
Social	• Innovative services as m Health, m Education and m Commerce • Satya Bharti School Programme • Mobile Application Tool for Enterprise (MATE) • Engaging with partners • A Marathon Effort • Expanding network to all 22 district in Jammu and Kashmir including Leh and Kargil	• 17% diminishment in client complaints when contrasted with earlier year • 35% lessening in accomplices' grievance calls as against a year ago • 24% expansion in number of students and 18% in number of teachers in the last two years at Satya Bharti Schools • 96% of parents with a girl child studying at Satya Bharti School needed her to pursue higher education (compared to 74% among parents whose child goes to other schools). • 94% of India-based suppliers with over 74% local procurement (in terms of value) • Connectivity in 22 regions of J&K–just private operator to give consistent integration in Kargil (J&K) • In 2012–13, INR 11.7 MN was raised through A MARATHON EFFORT initiative.
Environment	• Addressing Energy and Climate change • Gangaganj Solar power plant • Controlling Carbon footprints	• Achieved 15.8% decrease in CO_2 discharge/terabyte in the system base as against a year ago • Over 50 MN clients adjusted through e-bills, prompting sparing of around 21,400 trees annually • In 2012–13, more than 1,680 base stations were changed over from indoor to outdoor, disposing of the utilization of ventilation systems at these site areas. This try is far beyond Bharti Airtel's drive to convey 81% sites as outdoor. • In 2013, Airtel effectively finished the establishment of 25 biomass gasifiers, with a specific end goal to supplant DG at BTS locales, in Rajasthan circle.

		• The Indus Towers Green Sites venture, dispatched in 2011, was imagined to run telecom system operations without utilizing Diesel by enlarging power supply and other interchange energy initiatives without trading off on system network. As of March 2013, 9,000 of Indus Tower Airtel destinations have been effectively changed over into green locales by taking out the utilization of diesel. • Gangaganj is a 100 KWp non-infiltrating roof-top solar power plant and the first of its kind in the Indian telecom industry. This initiative has resulted in saving over 80 tons of CO_2 emissions. Airtel wants to imitate this initiative in its other MSC locations with 300KWp solar power plants.
Economical	•	• During FY 2012–13, the Company's aggregate spent on CSR is 0.58% of the net profit after taxes • Y-o-Y growth rate of 9% in gross revenues • Wireless subscriber base increased from 188 million to 196 million

4.3.2 *Social, Environment and Economic Impact of the CSR Initiatives*

Social

Airtel utilizes three vision columns to accomplish objective "Millions more are incorporated and enabled through maintainable social and financial improvement". Sustainable transformation in the society in order to overcome shortcomings in the health sector, employability opportunities and education in the main frame.

The use of telecom technology which is the core function of the company and using its various functionalities like Broadband platform, Digi-presence, Digital TV to be the path for improving accessibility to facilitate education, health and financial services..

Airtel expands horizon for development by involving people around them through various customer, employees and partner engagement initiative to build enduring relationship to integrate technology with people collaboratively helping society.

Environmental

Airtel consistently with the assistance of system base accomplices assess the effects of their business operations on any ecological debasement or harm by giving inventive arrangements that assistance in lessening negative effect on our planet.

Airtel demonstrates their responsibility towards vitality and environmental change, waste administration and activities to minimize carbon foot shaped impressions through different projects, occasions, classes, workshops and partner meets to lessen negative effects of base.

Airtel has spearheaded the idea of uninvolved foundation partaking in telecom over a decade ago.

Economic

There is a surge in the wireless customer base of the company in India region. Though a direct relation cannot be established between the corporate social initiatives taken by the company and the increase in gross revenue and customer base, the numbers suggest that the customer base increase can be indirectly attributed to the fact that the company has contributed 0.58% of its Profit to the CSR activities by most of the investment in the rural schools run by Bharti Foundation.

Again like other companies in the ICT domain, Bharti Airtel has also contributed heavily in the education segment. This creates a positive image in the eyes of the ultimate customer and hence it increases the brand valuation and market capitalization of the company. Though exact figures couldn't be found in the report but still increase in the gross revenue by 9% Y-o-Y and increase in the customer base by 8 million in India.

4.4 TATA Consultancy Services (TCS)

TCS is an IT services providing company. It provides business solutions to different industries globally. TCS has always worked in the direction to contribute as much as it can in the field of CSR and it has done efficient use of technology to maintain sustainable growth in social as well as environment sector. It has used environment friendly ways to resolve the industry problems and is also an example to other industries.

The main areas of TCS' CSR programs are:

- Education
- Health
- Environment.

4.4.1 *CSR Initiatives*

Area	CSR Initiatives	Impact
Social	Adult Literacy Program Advanced Computer Training Center (ACTC) Rural IT Quiz TCS maitree	202051 Adults Benefited Trained 136 Visually Impaired 14.54 Million Students Participated Rural Development
Environmental	Energy Management Paper Management Water Wastage Management Carbon Foot Print	39% Reduction in electricity consumption 78% reduction in paper consumption 13% reduction in fresh water consumption 34% reduction in carbon foot print
Economical		16.02% YoY Increase in revenue Reduction in Operating Cost Operating Margin Increased by 29%

4.4.2 *Social, Environmental and Economic Impact CSR Initiatives*

Social

TCS does a lot of CSR initiatives related to society in the fields of education, health, women empowerment and employment opportunities. In the initiative like Adult Literacy Program TCS is providing IT expertise to adults since 2000. Around 2,02,051 adults from rural areas experienced and learned about IT.

Under Advanced Computer Training Center TCS trained 136 Visually Impaired people and also provided employment to over 100 candidates across multiple companies.

TCS partnered with the Government of Karnataka and started and organised first of its kind Rural IT Quiz. The motto behind organizing such an event was to develop and instil knowledge of IT in the students of rural areas. The year 2014 saw a humongous participation of around 14.54 million from around 8000 schools.

Under its initiative TCS Maitree, TCS is working in the direction for improving and imparting education, environment and healthcare and providing economic help in rural India. During this initiative associates from TCS provided training to 45 women for making environment friendly jute bags as a 'Women Empowerment Initiative' also it developed the infrastructure for clean drinking water in the small villages due to which more than 570 children were benefitted and also they were provided guidance and support in the field of science and mathematics.

Initiative Insight seeks at instilling the concepts related to team work, effective communication with good presentation skills in addition to the technical knowledge. This project was undertaken in about 30 schools and covered 1433 students.

Operation Smile under the TCS banner an international level medical charity program for children's that worked on healing their smile and giving them a better life. TCS has tried to give its help in all the social fields like education, health and women empowerment in one or the other way.

Environmental

TCS views that an organisations sustainable growth is an imperative for its sustainable future. In this line TCS has contributed in all possible ways to help environment. It saw 39% reduction in per capita electricity consumption because of 550% increase in solar thermal installation and generating 2.3% of the required power from the renewable energy sources. TCS has worked in the direction of reducing the usage of paper as much as they can because it leads to tree cutting and also creates pollution. The success of this initiative can be gauged from the fact that TCS was able to achieve a reduction of 6.8% in its paper consumption reduction compared to the previous year and also achieving 78% reduction over the baseline.

By avoiding 1.2 million kl of water usage, by creating 217396 cum of rain water harvesting potential and by recycling 2.9 million kl of water TCS has significantly reduced the fresh water usage by 13%.

Also by Green building and Green IT initiatives TCS has been able to reduce the per capita carbon foot print by 34%. By these initiatives TCS has catered in all possible areas related to environment and has given its contribution for the betterment of environment.

Economic

Successful execution of sustainable growth of any company requires investment in capacity building, in human resource and also in new ventures. Due to the various initiatives done in the social and environmental domain by TCS its brand value has increased in the market which is reflected by the 16.02% YoY revenue increase, 17.4% growth in volume and gross margin and EBIT of 47.3% and 29.1% respectively.

Shareholder value is increased and company paid a dividend of ₹ 32 that is 37% of total profit. The operating cost reduced due to the initiatives like 550% increase in solar thermal installation, generating 2.3% of the required power from the renewable energy sources, 4 LEED certified campuses all these directly increased the profit margin and improved the financial performance.

5. FINDINGS AND CONCLUSION OF THE STUDY

The study was aimed at analysing some of the Corporate Social Responsibility initiatives taken up by Information, Communication and Technology (ICT) companies in India and its impact on social, environment and economic performance of the company using triple bottom line framework. Four companies were selected for study namely Infosys Limited, Wipro Technologies, Bharti Airtel and Tata Consultancy Limited. These four companies have been selected for this study as they are rated as top four companies as per CSR rating given by the research done by IIM Udaipur for top companies in CSR in different domains. As per the Research paper of IIM-Udaipur on Corporate Social responsibility, Infosys Limited topped the chart followed by Wipro Technologies Bharti Airtel and TCS, respectively.

After analysing the CSR initiatives of the above mentioned four ICT companies it can be concluded that the CSR initiatives of companies do have an impact on social, environment and economic performance of the company.

It was observed during the study that the contribution to sustainable technology can be penetrated in the market by big technological -companies which can offer the society a place where they can breathe fresh air, equip the potential talent with required resources, reducing the gender biased-ness, optimizing the best possible use of resources in the vicinity and reduce the carbon footprints so that a sustainable world with an efficient technology can be developed where the business flourishes along with the environmental factors.

The tableau of initiatives taken by the top four companies in the ICT domain gives us the glimpses of the responsibilities these companies have shouldered from time to time for the society and complying with the Companies act, 2013.

In the next 2–3 years of time span *i.e.* by 2017–18 we expect these companies to be completely dependent on the Renewable resources. We expect them to create an ecosystem which is driven completely on renewable resource platform by reducing GHG emissions, reducing carbon intensity, reducing water intensity and electricity intensity.

The commonalities in the companies chosen can be chalked out with respect to the technological advancements these companies have taken in their businesses for reducing the operational cost and hence increasing the profit margins for the company. Also, all the companies have increased their revenues in the financial year 2012–13 which is indirectly related to the fact that these companies have reduced their dependencies on non-renewable resources by swapping it with efficient renewable resources. As we have seen in case of these companies who have reduced the paper work by adopting the digitalizing technology in the company and hence reducing the cost of the paper and also reducing dependency on paper work.

Moreover, these companies have been realizing the talent available in the rural areas and they are capitalizing on these people by providing educational infrastructure like schools, libraries, internship opportunities etc.

A roadmap to achieve business goals and sustainability goals can be achieved by these companies by 2017–18 financial year, by following more aggressive and assertive biodiversity, human rights, educational and green deployment of resources.

More of such practices need to be followed by companies in the ICT domain as these have the potential to bring technological change. The ICT industry growing at a substantial rate of 13–15% and since the sector is growing at a good pace and is expected to grow in the near future we expect a rise in these socially viable and Environmental friendly initiatives from other companies as well.

REFERENCES

Bhattacharya, C.B. and Sankar Sen (2003). "Consumer-company identification: A framework for understanding consumers' relationships with companies," *Journal of Marketing,* 67(2), 76–88.

Brown, Tom J. and Peter A. Dacin (1997). "The Company and the Product: Corporate Associations and Consumer Product Responses," *Journal of Marketing*, 61 (January), 68–84.

Creyer, Elizabeth H. and William T. Ross (1997). "The Influence of Firm Behavior on Purchase Intention: Do Consumers Really Care About Business Ethics?" *Journal of Consumer Marketing*, 14 (6), 421–32.

Ellen, Pam Scholder, Lois A. Mohr, and Deborah J. Webb (2000), "Charitable Programs and the Retailer: Do they Mix?" Journal of Retailing, 76(3), 393–406.

Kim, Kyungjin and Jongchul Park (2009), "The effects of the perceived motivation type toward Corporate Social Responsibility activities on consumer loyalty," *Journal of Global Academy of Marketing Science*, 19 (3), 5–16.

Margarita Tsoutsoura (2004), "Corporate Social Responsibility and Financial Performance" Page No. 6.

McWilliams, A. and Siegel, D. (2000) "Corporate social responsibility and financial performance: Correlation or misspecification?" *Strategic Management Journal*, 21(5): 603–609.

Mullen, J., 1997. Performance-based corporate philanthropy: how 'giving smart' can further corporate goals. *Public Relations Quarterly*, 42(2), pp. 42–48.

Organ, Dennis (1988), *Organizational Citizenship Behavior: The Good Soldier Syndrome,* D. C. Health, Lexington, MA.

http://www.wbcsd.org/work-program/business-role/previous-work/corporate-social-responsibility. aspx viewed on 16th July, 2015.

http://www.unido.org/en/what-we-do/trade/csr/what-is-csr.html viewed on 16th July, 2015.

Vidhi-Bhargava-and-Divya-Shikha_Sustainability-Practices-at-Infosys – Research Paper on Infosys.

http://www.infosys.com/sustainability/Documents/infosys-sustainability-report-2012-13.pdf viewed on 19th July, 2015.

http://www.tcs.com/about/corp_responsibility/cs-report/Documents/GRI-2013-Sustainability-Report-271014.pdf; viewed on 18th July, 2015.

Top_indian_Companies_for_CSR_2014_Report- Research report by IIM-Udaipur.

http://www.airtel.in/sustainability-file/common/files/Sustainability_Report-FY_2012-13.pdf viewed on 15th July, 2015.

http://www.airtel.in/wps/wcm/connect/f5443281-e006-4f29-938d-46a4fc2ea67c/Annual-Report-of-Bharti-Airtel-for-FY-2012-2013.pdf?MOD=AJPERES viewed on 22nd July, 2015.

http://www.wipro.com/documents/investors/pdf-files/Wipro-annual-report-2012-13.pdf viewed on 18th July, 2015.

https://www.wipro.com/documents/FINAL_Sustainability_Report_IT_270312_6mb.pdf viewed on 20th July, 2015.

http://www.sciencedirect.com/science/article/pii/S0970389610000376 - 05/08/2015 viewed on 5th August, 2015.

Open Innovation—A Need of the Hour for Indian Small and Medium Enterprises

Sumukh S. Hungund[1] and K.B. Kiran[2]
School of Management, National Institute of Technology, Karnataka–575 025, India
E-mail: [1]sumukh.hm13f03@nitk.edu.in; [2]kbheem@nitk.ac.in

ABSTRACT: *Innovation is vital to sustain a cutthroat advantage in the competitive business environment and to achieve leadership. Ideas are the most essential input for an innovation process to start. Innovation has been considered as a prominent engine of growth. Also, Innovation plays a pivotal role in providing sustainability for the firm. But yet firms are not clear about the type of innovation approach that needs to be adopted for idea generation and product development. A condensed product life cycle, increasing research and development cost, rapid information exchanges, and increasingly inter-connected customers have made a shift towards approach of openness for innovation activity. A firm needs to choose between Open Innovation practices and Closed Inno-vation practices for its sustainable development. Small and Medium Enterprises (SME) of Information Technology Industry in India have adopted innovation practices to the tune of fifty seven percent. Open Innovation has emerged as hottest topics for research in the innovation management space. Open Innovation practices is considered a key tactical approach to enhance the commercialization of innovations among firms. Studies related to Open Innovation are connected with large firms and in the context of West. However, in the context of India, a limited work related to Open Innovation practices and SME can be seen. In addition, a large numbers of issues with respect to Open Innovation Theory are not clear. Hence, the current paper critically reviews the existing literature and devel-ops a conceptual framework to establish a relationship between Firms, Open Innovation Practices, SME Characteristics, and firm performance. The paper establishes a need to find out the Open Innovation practices among SME segments of the technology oriented firms in India. The paper also presents the research questions and research objectives of the study along with hypotheses. The paper explains the research methodology adopted to meet the objectives of the study. The paper concludes with the need of research and the contribution that will be made from this study to the world of academia.*

Keywords: Open Innovation, Closed Innovation, In-bound Open Innovation, Outbound Open Innovation and SME.

1. INTRODUCTION

Innovation has been widely accepted as an essential competitive tool for any enterprise for a sustainable growth (Drucker, 1985). Innovation practices have been considered as a prominent growth engine by Large and Small and Medium Enterprises (SMEs) (Yifeng, 2011; Mashilo and Iyamu, 2012). National Knowledge Commission report (2007) reveals that innovation has the most significant impact on competitiveness for large firms while for SME's innovation will make indelible impact on increase in market share. But SMEs said to have difficulty in implementing innovation practices (Iakovleva, 2013). Today, the innovation process is undergoing profound changes in the way it is managed (Chesbrough, 2003). The studies related to Open Innovation are gaining more importance in today's context (Wang and Tang, 2013). The Open Innovation practices have been displayed by the technology intensive firms (Mazini *et al.,* 2013). Chesbrough (2003) defines Open Innovation as "the purposeful exchange of information or idea to grow innovation activity, and expand the business to commercialize the innovation, respectively". Also Open Innovation models stresses on exploiting a wide source of information to do innovation (West and Gallagher, 2006). Also it is said that Open Innovation approach enhances firm's prospect to achieve business growth by sprouting new products to market (Freel, 2006). Given the background of Innovation's significance and recent trends of Open Innovation studies across the globe it is of utmost importance to find out the extent to which the Open Innovation practices have been adopted among technological SMEs in India. Hence this paper focus of the need for Open Innovation practices among software product SMEs. The paper is divided into following sections. The paper begins with introduction followed by review of literature, conceptual framework, research questions and objectives, hypotheses statement, results and discussion and then conclusion and limitations of the study.

2. REVIEW OF LITERATURE

This section reviews the literature on Open Innovation practices adopted and also on Open Innovation and firm performance.

Lichtenthaler (2008) reveals that several firms follow Closed Innovation practices to innovation despite an inclination toward Open Innovation. Also opens that there is a need to find out the Open Innovation practices adoption among small firms. V. van de Vrande *et al.* (2009) finds out medium size firms are more active in engaging practices of Open Innovation than firms of smaller size in Dutch. Also, opines that there is need to study Open Innovation adoption and practices in broader samples across different geographies. Open Innovation leads to business growth (Huang *et al.,* 2010) Jayawardhana and Surangi (2010) opine that a considerable variation is noted among medium and small firms during the process of adoption of Open Innovation approach. Also, reveals firm's growth and sustainability is attributed to the embracement of practices of Open Innovation. Gumus and Cubukcu (2011) opine that awareness of Open Innovation among Turkish firms is very low and mentions that for a sustainable growth a culture of innovation is essential in the firm. Also opines firm's characteristics are not related to innovation practices adopted by firm. Xin and Wang (2011) reveal that SMEs needs Open Innovation for sustaining rather than for transformation to large organization. Also mentions that practices of innovation should be carefully adopted by SMEs. But feels that the type of Open Innovation practices considered for firm performance is unexplored. Xu and Zheng (2012) in their work discuss about definition, background and research foundations of open innovation and suggests about the need to study factors influencing open innovation. Huizingh (2010) opines that there are many concepts of Open Innovation which needs to be understood and feels that there is lack of its understanding among SMEs. Kafouros and Forsan (2012) suggest that university collaboration with firm need to be explored and an integrated approach of Open Innovation and firms' performance needs to be investigated. Tian and Feng (2010) investigated the types of outside sources of technology in Open Innovation and finds that the sources of technology transfer include suppliers, customers, academic institutions, research institutes, competitors, and large firms. Abulrub and Lee (2012) opinioned that company size and market type influences to adopt open innovation practices. Further feels that external partners are very important for firms to adopt open innovation. Since the study considers both large and small companies, the results need to be investigated only for SMEs. Balasubrahmanya M.H. (2012) SMEs internal technical competence and their nature of innovation help them to fetch external support. Further felt that SMEs technical competency clubbed with external support exploit market opportunities to achieve higher innovative performance. Further suggests that there is a need to study the type external support need for SMEs in the Indian context. Lukas *et al.* (2012) reveals that successful innovation for a firm can be achieved through collaborative approaches. Janeiro *et al.* (2013) finds that successful firms depend on universities or academic institutes for innovation. Further opines that fundamental associations is their between firm's innovation and access to sources. The study suggest that there is a need to study the reasons that encourages the firms to look out for outside partners and how access to knowledge that is external to firm shapes and impacts innovation. Rangus and Drnovsek (2013) opine that the most common practices of Open Innovation are customer participation, employee's participation and pre-venturing. The study reveals that innovation activity of firms depends on collaboration with suppliers and also with customers. Also, the study opines that smaller firms are inclined towards selling or licensing of their Intellectual Property. Ades *et al.* (2013) analyses three case firms whose innovation management processes have been fused and found that the firm culture as an obstacle to adopt practices of Open Innovation. Segers (2013) observed that there is a strong bonding between collaboration partners like research establishments, academic institutions, existing large companies, and firms of Biotechnological segment in Belgium. Also, feels that innovation happens across university-based firms. Further mentions that, there is a need to examine the Open Innovation practices and the observation about collaboration also needs to examined and validated across high technology based firms. Revutska (2013) feels that Open Innovation business model is a viewpoint from strategic development. Companies benefit from the quick commercialization of their ideas and will be able to improve their experience through the diffusion of innovations, among other companies in the market. Further acknowledges the role played by academic institutions in the process of creating Open Innovation models. These institutions may be engaged in the creation and commercialization of knowledge. Deegahawature (2014) suggest that firms implement inbound open innovation at a moderate level and suggest that firms that adopt inbound open innovation should be cautious on capabilities and environment turbulence. The study fails to explain about technology exploration through external agents like academia. Hence there is a need to study the collaboration activities and its influence. The literature review suggests that the studies conducted so far in view of Open Innovation are largely in the context of the West. However there are limited studies which compare both innovation practices. The prior empirical studies have only concentrated on adoption of Open Innovation only. Very few studies discuss about Open Innovation practices and firm performance (Mazzola *et al.,* 2012; Cozzarin, 2004 and Santos *et al.,* 2014) but these studies are in the context of European and American firms. Also there is limited evidence on adoption of Open Innovation approach and performance of firm (Sisodiya *et al.,* 2013). Hence there is a definitive need to study different Open Innovation approach and influence of this approach on performance of firm.

3. CONCEPTUAL FRAMEWORK FOR THE STUDY

Firm Performance which is dependent variable is measured through change in Market share. The Open Innovation practice which is the independent variable is measured through the extent to which the following practices are adopted by the software firms. The practices include Collaboration with academic institution or universities, suppliers, customers, and R&D labs, the spin-offs made by the organizations as teams of product development or as separate entity, Alliances made with other organization and licensing of Intellectual property.

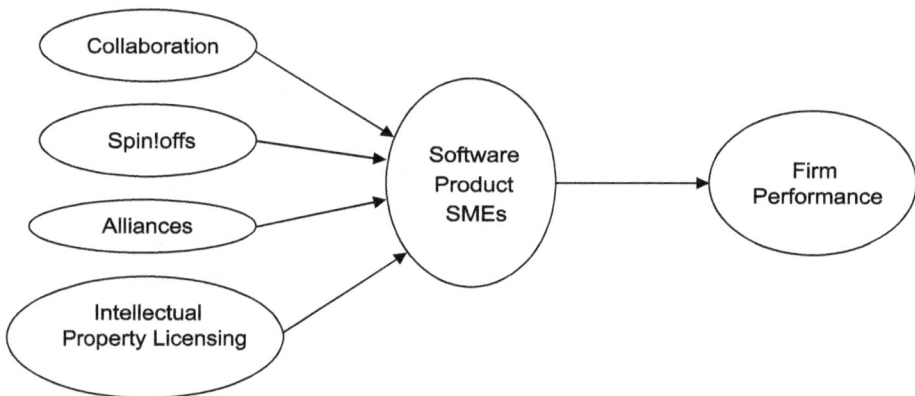

Fig. 1: Conceptual Framework of the Study

Source: Literature review.

3.1 Research Questions

1. What is the status of awareness and adoption of Open Innovation practices among Software SMEs?
2. Do Open Innovation practices influence the performance of firm?

3.2 Research Objectives

1. To assess the awareness level and adoption level of Open innovation.
2. To examine whether the Open Innovation practices influences firms performance.

3.3 Hypothesis Statement

To measure the research objectives following hypotheses is stated. H_{1a} and H_{2a} measure the research objective 1 and H_{3a} measure the research objective 2:

- H_{1a}: The awareness level of Open Innovation practices is significant among firms.
- H_{2a}: The adoption level of Open Innovation practices is significant among firms.
- H_{3a}: Open Innovation practices do influence firm performance.

4. RESEARCH METHODOLOGY

Primary data of the study is gathered through on a structured questionnaire. A google doc has been developed and the survey questionnaire link is sent to the CEOs/CTOs/VPs/Product heads of various Information Technology product firms whose headcount is less than 250 through an e-mail. The sample has an appropriate mix of core product companies, product and services companies and product as service companies. The survey link has been sent to 40 companies and a completed response is received from 30 companies with a response rate of 75%.

4.1 Dependent Variable and Independent Variable

In the current study, the dependent variable is firm performance and is measured through change in market share. A multiple regression is performed to measure the firm performance.

The key independent variable is Open Innovation practices which consists practices such as Collaboration, Spin-offs, Alliances, Intellectual Property Licensing.

5. RESULTS AND DISCUSSION

The reliability of the items of questionnaire is measured by conducting a reliability test for all items which are on ordinal scale. The Cronbach's α is found to be 0.742 for 47 items on ordinal scale. The 6 items which discusses firm performance, the Cronbach's α is found to be 0.683 and the items which measures Open Innovation approaches and practices, the Cronbach's α is found to be 0.725 . All the values are found to be acceptable.

The awareness of the term Open Innovation is only 46.7%. This indicates that the term has a considerable awareness. From one-sample test, it is very much evident that among the firms there is a significant awareness of the term Open Innovation. The outcome of the One-Sample Test has been presented in Table 1. From the Table 1 it is observed that the t value is 5.037 and is significant at 95% confidence interval. This indicates that statistically H_{1a} is accepted. This means the awareness of the term Open Innovation among firms is a significant. Whereas the adoption level of the Open Innovation practices among firms is only 43%. From a one-sample test, it is evident that the firms are willing to adopt or have adopted Open Innovation practices. From the Table 2 it is observed that t value for adoption of Open Innovation is 4.709. The t value is statistically significant at 95% confidence level. Hence H_{2a} is accepted is accepted. This indicates that firms among the sample are willing to adopt Open Innovation practices.

Table 1: One Sample Test on Awareness of OI

	Awareness of OI		*t*	Degrees of Freedom	Sig. (2-tailed)
	Yes	No			
Aware of the Term "Open Innovation"	46.7%.	53.3%	5.037	29	0.000

Source: Sample survey.

Table 2: One Sample Test on Adoption of OI

	Adoption of OI		*t*	Degrees of Freedom	Sig. (2-tailed)
	Yes	No			
Adoption of Open Innovation practices	43%	57%	4.709	29	0.000

Source: Sample survey.

The Table 3 represents a regression model of the firm performance and Open Innovation practices. The Table 3 it can be said that the predictor variables *i.e.* independents variables has a good relation with the dependent variable and the model is also significant at 95% confidence interval. Also from the Table 4 it is clear that collaborations with the supplier and Universities are the most adopted practice among the Open Innovation practices and these practices have considerable influence on firm performance. It could be seen that the t value for the collaboration with supplier is significant at 95% and collaboration with universities is significant at 90%. Also indicates, collaboration with academic institutions may influence the firm performance. Since the f-value and values of Durbin-Watson is found to be acceptable at 95% confidence level, the hypothesis H_{3a} is not rejected.

Table 3: Summary of Descriptive Regression

R	R^2	Adjusted R^2	Std. Error	F value	Degrees of Freedom	Significance	Durbin-Watson
0.673	0.453	0.279	0.989	2.603	7	.041	2.012

Source: Sample survey.

Table 4: Summary of Coefficients

Independent Variables	Beta	Standard Error	t-Value	Significance Level
Constant	6.382	1.306		0.000
Collaboration with Universities	0.403	0.207	1.946	0.065
Collaboration with Suppliers	0.475	0.229	2.078	0.050
Collaboration with R&D labs	−0.172	0.260	−0.660	0.516
Collaboration with Customers	−0.378	0.255	−1.480	0.153
Licensing idea/technology IPR to Partners	−0.196	0.184	−1.068	0.297
Alliance for New Product Development	−0.303	0.231	−1.312	0.203
Spin-off my Product Team to Develop a Product	0.159	0.186	0.856	0.401

Source: Sample survey.

6. LIMITATION AND CONCLUSION

The study infers that the notion of Open Innovation is very new among the Indian firms particularly in the small and medium segments. But still organizations are open to new practices. Even though the term might be new but the practices are being followed. Also the results indicate that Open Innovation practices help the firms to improve their performance. Collaboration is a key Open Innovation practice that firms have adopted extensively for firm performance. The collaboration with suppliers and academic institution are preferred compared to customers and R&D labs. The concept of Spin-offs and Intellectual Property is still new. The results on a large sample may differ and needs to be explored. Thus the results of pilot study indicate that there is a definitive need to study the Open Innovation practices among small and medium firms. The results indicate that these practices if adopted could also help the firm to grow to the next level. The limitation of the current study is a pilot study. The study is limited to the companies located in the Bangalore Ecosystem.

REFERENCES

Abulrub, A. and Lee, J. (2012). Open innovation management: Challenges and prospects. *Procedia—Social and Behavioral Sciences, 41,* 130–138.

Ades *et al.* (2013). Implementing Open Innovation: The case of Natura, IBM and Siemens. *Journal of Technology Management and Innovation,* 8, 12–24.

Arip, B. and Romy, L. (2012). Business Process Reengineering in Motorcycle Workshop X for Business Sustainability, *Procedia Economics and Finance,* 4, 33–43.

Balandin, S. (2010). Experience and Vision of Open Innovations in Russia and Baltic Region: The FRUCT Program. *IEEE,* 8, 5–10.

Balasubrahmanya, M.H. (2012). External Support and Innovation Performance of MEs in Bangalore: Role of Firm Level Factors. *IEEE,* 65–70.

Chesbrough, H., Vanhaverbeke, W. and West, J. (2006). Open Innovation Researching a new Paradigm. *Oxford university Press,* 1–27.

Chesbrough, H. (2003). The era of open innovation. *Sloan Management Review,* 44(3): 35–41.

Chesbrough, H. (2004). Open Innovation: The new imperative for creating and profiting from Technology. *Harvard Business Review,* 10–20.

Deegahawature (2014). Capabilities and Implementation of Inbound Open Innovation: Evidence from LMT Firms in Technologically Less Advanced Countries. *European Journal of Business and Management,* 6(7), 286–295.

Drucker, P. (1985). Innovation and Entrepreneurship: Practice and principles. *London: Heinemann,* 1–6.

Gumus, B. and Cubukcu, A. (2011). Open Innovation Survey in Top Turkish Companies. *IEEE,* 1–6.

Gunday, G. *et al.* (2011). Effects of innovation types on firm performance. *International Journal of Production Economics,* 133, 662–676.

Hakkim, R. and Heidrick, T. (2008). Open Innovation in the Energy Sector, *IEEE,* 565–571.

Huang, T., Wang, W., Yun, K., Tseng, C. and Lee, C. (2010). Managing technology transfer in open innovation: The case study in Taiwan. *Modern Applied Science,* 4(1), 2–11.

Huizingh, E. (2010). Open innovation: State of the art and future perspectives. *Technovation,* 1–8.

Iakovleva, T. (2013). Open Innovation at the Root of Entrepreneurial Strategy: A Case from the Norwegian Oil Industry. *Technology Innovation Management Review,* 17–22.

Janerio, P., Proenca, I. and Goncalves, V. (2013). Open innovation: Factors explaining universities as service firm innovation sources. *Journal of Business Research,* 2017–2023.

Jayawardhana, A. and Surangi, H. (2010). Open Innovation practices in women owned handicraft manufacturing SMEs: A case from central province, Sri Lanka. *In: ICBI,* 1–16.

Jianzhong, L. (2010). Research on the Mechanism of Hi-tech Enterprises' Open Innovation on the Basis of Knowledge Chain. *IEEE,* 163–167.

Jonthan, Y. and Athreyi, K. (2013). Exploring innovation through open networks: A review and initial research questions. *Management Review,* 25: 69–82.

Kafouros, M. and Forsan, N. (2012). The role of open innovation in emerging economies: Do companies profit from the scientific knowledge of others? *Journal of World Business,* 47, 362–370.

Kolaskar, A., Anand, S. and Goswami, A. (2007). Innovation in India. *National Knowledge Commission,* 7–31.

Lichtenthaler, U. (2008). Open Innovation in Practice: An Analysis of Strategic Approaches to Technology Transactions. *IEEE transactions on engineering management,* 55(1), 148–157.

Lukac, D., Rihter, J.D., Rogic, M. and Mikela, M. (2012). Open Innovation Model in the ICT Industry – The Case of the German Telekom, *IEEE*, 48–51.

Mashilo, M. and Iyamu, I. (2012). The Openness of the Concept of Technology Open Innovation, *IEEE*, 487–492.

Mazinia, S.R. *et al.* (2013). Open innovation and user's involvement in new product development: A case study in the automotive sector. *Product: Management and Development*, 11, 49–55.

Munkongsujarit, S. and Srivannaboon, S. (2011). Key Success Factors for Open Innovation Intermediaries for SMEs: A Case Study of iTAP in Thailand. *IEEE*, 1–10.

Rangus, K. and Drnovsek, M. (2013). Open Innovation in Slovenia: A comparative analysis of different firm size. Economic and Business Review, 15(3), 175–196.

Revutska, N. (2013). Open innovation as a strategic model of modern business. *European Scientific Journal*, 1, 1857–7881.

Rothwell, R. (1992). Successful industrial innovation: Critical factors for the 1990s. *R&D Management*, 22(3), 221–239.

Segers, P. (2013). Strategic Partnerships and Open Innovation in the Biotechnology Industry in Belgium. *Technology Innovation Management Review*, 23–28.

Sun, X. and Wang, Q. (2011). Open Innovation in Small and Medium Enterprise under the view of Knowledge Management. *IEEE*, 4690–4693.

Susanne, D. and Pirjo, S. (2013). Success Factors of Open Innovation—A Literature Review. *International Journal of Business Research and Management*, 4(4), 111–131.

Tian, D. and Feng, Y. (2010). The Categories of External Technology Sources in Open Innovation. *IEEE*, 1–4.

Vanhaverbeke, W. (2013). Rethinking Open Innovation beyond the Innovation Funnel. *Technology Innovation Management Review*, 6–10.

Venturini and Verbano (2013). Openness and innovation: an empirical analysis in firms located in the Republic of San Marino. *International Journal of Engineering, Science and Technology*, 5(4), 60–70.

Vrandea, V. *et al.* (2009). Open Innovation in SMEs: Trends, motives and management challenges. *Technovation*, 29, 423–437.

Wang, W. and Tang, J. (2013). Mapping Development of Open Innovation Visually and Quantitatively: A Method of Bibliometrics Analysis, *Asian Social Science*, 9(11), 254–269.

West, J. and Gallagher, S. (2006). Challenges of open innovation: the paradox of firm investment in open-source software. *R&D Management*, 36(3), 319–331.

Xu, Y. and Zheng, J. (2012). Open Innovation Literature Review and Outlook. *Proceedings of the IEEE ISMOT*, 558–562.

Yifeng, Xie (2011). Open Innovation of small and medium sized enterprises and R&D Public Services Platform: A case about the VIC model of Zhangjiang Hi-tech Park, *IEEE*, 1–5.

Purchase Intention Towards Green Products of Younger Consumers in India

Vimal Chandra Verma and Devashish Das Gupta
[1]Uttar Pradesh Technical University, Lucknow, India
[2]Department of Marketing Management, IIM, Lucknow, India
E-mail: vimalchandraverma@gmail.com; [2]devashish@iiml.ac.in

ABSTRACT

Aim/Purpose: *Young consumer market is becoming more attractive for the companies offering environmental friendly products due to their awareness and income. The purpose of the study was to investigate antecedents and factors behind purchase of green products.*

This study seeks to explore young consumers' intention towards buying eco-friendly products.

Methodology: *In this paper, a quantitative methodological approach is adopted and data are collected using a questionnaire, sample size was (n = 160) young people of different technical institutions of Lucknow city. The questionnaire is analyzed using Statistical Package for Social Science (SPSS) version 20.*

Findings: *The consumers feel that green products are useful for the health. People who think that green products are useful are also buying green products. Education plays a crucial role behind the purchase of green products.*

Gender wise there are no difference towards the selection of T.V programs devoted to environmental topics and issues. Education was found as differentiator while selection of T.V. programs devoted to environmental friendly products. Mostly respondents agreed that they learned from their parents about usefulness of green products and accepted that it should be purchased. Media plays a crucial role in influencing consumers towards the purchase green products.

Originality/Value: *This paper fulfils a need for information on younger consumers' intention towards buying green products and offers advice to marketers attempting to better understand the emerging and increasingly important consumer segment.*

Managerial Implications: *Young consumers' attitude towards products always has crucial issue for a marketer due to their number and India is kwon as a country having highest number of young consumer globally. Marketers will be in position to better prepare different marketing strategies and stimuli for marketable product development and its implication on the basis of the consumer's preferences. Findings of this study about different factors that have impact on younger consumers' attitude towards buying eco friendly products will certainly provide some new insights for rethinking and to go deeper towards exploring green marketing.*

Limitations: *The survey for the study was limited to Lucknow city only. Further a more comprehensive survey can be done taking specific green products like view towards LED etc.*

Keywords: Intentions, Purchase, Green Products, Young Consumers.

1. INTRODUCTION

Over the past few years, concern towards environment has become vital area of interest for companies and this is due to consumers, attitude to save environment. This has led to increase in demand for green products has grown significantly all around the world and India is also no exception due to consumers increased interest in conserving energy and saving environment. For the past few years large number of people felt that they are highly concerned about the environment and problems related to environment (Diekmann and Franzen, 1999; Dunlap and Mertig, 1995). Consumers have become more environmental conscious and started buying green products. The concern towards environmental decay has given rise to a new group of consumers such as green consumers. These consumers can be said as one who prefers to purchase products which do not harm the environment during production and use. More ever due to this drastic change it has become crucial for the business enterprises to study this consumer segment. Green marketing can be broadly defined as selling products and providing services based on environmental concern and benefit. Green marketing has been found as a major trend in current business scenario (Kassaye, 2001). According to Ottman, 1992 green marketing has been found as strategic marketing approach in recent business. Customers are therefore become more sensitive in their preference to choose and buying the products (Sarigollu, 2009). The demand of eco-friendly products is increasing day by day. Many consumers are aware regarding various environmental issues and prefer to select products that less harmful for the environment. It can be traced from various studies that the growth of environment friendly products is increasing due to environmental degradation.

2. LITERATURE REVIEW

Mostafa (2007) defined green purchase behavior as the consumption and use of products that benefits the environment as well as also recyclable or responsive and sensitive to ecological concerns. Consumer's attitude and belief are two vital factors that would influence their green food consumption decisions. Roberts (1996) found in his study that the consumers who were environmentally conscious and had faith the certain ecological activities can improve environment were highly liked to perform green consumer behavior. Novera Ansar (2013) said in his study that age and education are associated with environmental literacy and also concluded that advertisement focused on green marketing, price and packaging have positive relation with consumer's intention to buy and use green products.

2.1 Environmental Awareness and Consumption of Green Products

Having awareness towards environmental problems is primary step for the consumers to know and understand the different forms of behavior towards environment including green purchasing behavior. Attitude and this awareness have a positive influence on their purchase behavior.

2.2 Education and Green and Green Products

Hustad and Pessemier (1973) revealed in their study that education level is a vital antecedent for consumers to be concerned with different environmental issues. N. Mahesh (2013) found in his study that consumers who had higher education showed more perceived value than those who have low education and more educated people were found more likely to purchase green products in comparison to the people who had low education.

3. HYPOTHESIS

- Gender effect the user pattern of green product.
- Education level affects the user pattern of green product.
- Age effect the user pattern of green product.
- People who think green product are useful also purchasing the green product.
- People who are highly educated are purchasing environmental friendly product more than the less educated people.

4. SAMPLE PROFILE

The Survey was conducted among the student to know the purchase intention of green products like green tea, energy saving Bulb, eco friendly card *etc.* The Survey was conducted among the Male and female where 62% of the respondent were male lies in Male category rest of 38% respondent were female.

Table 1: Gender

	Frequency	Percent	Valid Percent	Cumulative Percent
Male	100	62.5	62.5	62.5
Female	60	37.5	37.5	100.0
Total	**160**	**100.0**	**100.0**	

Table 2: Education

	Frequency	Percent	Valid Percent	Cumulative Percent
Bachelor Degree	50	31.3	31.3	31.3
Master's Program	110	68.8	68.8	100.0
Total	**160**	**100.0**	**100.0**	

Table 3: Age

	Frequency	Percent	Valid Percent	Cumulative Percent
19–21 Years	126	78.8	78.8	78.8
22–26 Years	24	15.0	15.0	93.8
27–30 Years	2	1.3	1.3	95.0
More than 30 Years	8	5.0	5.0	100.0
Total	**160**	**100.0**	**100.0**	

The covered sample was also categorized in two education level Bachelor segment and Master's Program. Where 31% respondents were holding bachelor degree and rest of all almost 69% respondent were having Master's degree.

Same way if we categorized the covered sample in age groups then we can distribute the respondent in different age groups brackets. Where 79% respondent were in of 19–23 Years age bracket, 15% respondent were in 22–26 years age group brackets, rest of the 6% respondent were lies in 27–30 years and more than 30 Years of age groups brackets.

H1: Gender effects the user pattern of green products.

Table 4: Chi-Square Tests

	Value	df	Asymp. Sig. (2-sided)
Pearson Chi-Square	12.205[a]	5	.032
Likelihood Ratio	14.218	5	.014
Linear-by-Linear Association	3.386	1	.066
N of Valid Cases	160		

a. 6 cells (50.0%) have expected count less than 5. The minimum expected count is 1.50.

The above table shows that respondent of different gender are saying green product are use full for the health among the sample respondent most of the people are saying green product are useful for health. The Chi square table also supports the same statement with significance value of 0.032 that is below 0.05 means there is significance difference between the both gender group male and female.

H2: Education level affects the user pattern of green product.

Table 5: Chi-Square Tests

	Value	df	Asymp. Sig. (2-sided)
Pearson Chi-Square	13.162[a]	5	.022
Likelihood Ratio	13.991	5	.016
Linear-by-Linear Association	1.579	1	.209
N of Valid Cases	160		

a. 6 cells (50.0%) have expected count less than 5. The minimum expected count is 1.25.

The above table shows that respondent of different education bracket are saying green product are use full for the health among the sample respondent most of the people are saying green product are useful for health. The Chi square table also supports the same statement with significance value of 0.022 that is below 0.05 means there is significance difference between education groups.

H3: Age effect the user pattern of green product.

Table 6: Chi-Square Tests

	Value	*df*	*Asymp. Sig. (2-sided)*
Pearson Chi-Square	31.174[a]	15	.008
Likelihood Ratio	28.743	15	.017
Linear-by-Linear Association	.969	1	.325
N of Valid Cases	160		

a. 19 cells (79.2%) have expected count less than 5. The minimum expected count is .05.

The above table shows that respondent of different age group people are saying green product are use full for the health among the sample respondent most of the people are saying green product are useful for health. The Chi square table also supports the same statement with significance value of 0.08 that is below 0.05 means there is significance difference between age group brackets.

H4: People who think green product are useful also purchasing the green product.

Table 7: Descriptive Statistics

	Mean	*Std. Deviation*	*N*
Green Products are Useful for Health	6.05	1.154	160
Often Buy Green Products	5.71	1.178	160

Table 8: Correlations

		Green products are Beneficial for Health	*Often Buy Green Products*
Green Products are Useful for Health	Pearson Correlation	1	.325[**]
	Sig. (2–tailed)		.000
	N	160	160
Often Buy Green Products	Pearson Correlation	.325[**]	1
	Sig. (2–tailed)	.000	
	N	160	160

**. Correlation is significant at the 0.01 level (2–tailed).

The table shows that people who are thinking green product are useful for health also buying the green product. Here the correlation value among the both parameter is positive 0.325 it means that those people who are thinking green product are useful also buying the green product. Hence the hypothesis is accepted.

H5: People who are highly educated are purchasing environmental friendly product more than the less educated people.

Table 9: Descriptive Statistics

	Mean	Std. Deviation	N
Green Products are Useful for Health	6.05	1.154	160
I Intend to Purchase Green Products in Future	6.40	.906	160

Table 10: Correlations

		Green Products are Useful for Health	I Intend to Purchase Green Products in Future
Green Products are Useful for Health	Pearson Correlation	1	.330**
	Sig. (2–tailed)		.000
	N	160	160
I Intend to Purchase Green Products in Future	Pearson Correlation	.330**	1
	Sig. (2–tailed)	.000	
	N	160	160

** Correlation is significant at the 0.01 level (2–tailed).

The above table shows that people who are thinking green product are useful for health also intended to purchase the green product. Here the correlation value among the both parameter is positive *i.e.* 0.330 it means that those people who are educated use green product intended to purchase the green product. Hence the hypothesis is accepted.

5. DISCUSSION AND FINDINGS

The analysis shows that consumers perceive that green products are useful for the health and Chi square table also supports the same statement with significance value of 0.032 that is below 0.05 means there is significance difference between the both gender group male and female. Age was also found to have significant impact on purchase decisions of green products. More ever consumers who thought green products are good for health also buying green products. The people with higher education preferred to go for buying green products since there is positive correlation between both the parameters.

6. LIMITATION

This study was basically done on students of different engineering and management colleges of luck now. The students of other institutes of India in highly urban locations may have an effect on their perspective on environment. Therefore, the findings of this study may not be able to generalize to whole younger population of India.

7. CONCLUSION

The evidence suggests that education has a positive impact in consumers' preference while going for green products. Since the study was basically conducted on younger consumers of India it showed their high interest towards using eco friendly products. Gender also played a crucial role while taking green buying decisions.

8. MANAGERIAL IMPLICATIONS

The main purpose of this research study was to explore and get a deeper insight about the relationship between young consumers' intention towards green products. An analysis of the coefficients of each dimensions such as gender, education and age showed positive relationship towards buying green products. A majority of respondents believed that green products are useful for the health. The consumers were also found to have purchasing green products who had perception that it is good for the environment.

The findings of the study also highlighted that while designing their offering the manufactures should take into consideration health as a key factor. Participants in the study had strong and consistent attitude towards environmental concern.

REFERENCES

Ansar, N. (2013). Impact of Green Marketing on Consumer Purchase Intention. MJSS.

Dunlop, R. and Mertig, A. (1995). Global Concern for the Environment: Is Affluence a prerequisite? Journal of Social Issues, 51(4), pp. 121–137.

Diekmann, A. and Frazen, A. (1999). The Wealth of Nations and Environmental Concern. Environment and Behaviour, 31(4), pp. 540–549.

Hustad, T.P. and Pessemier, E.A. (1973). Will the real consumer activist please stand up: An examination of consumers' opinions about marketing practices. *Journal of Marketing Research*, 319–324.

Mostafa, M. (2007). A hierarchical analysis of the green consciousness of the Egyptian consumer. *Psychology and Marketing*, 24(5), pp. 445–473.

Ottman, J. (1992). Industry's Response to Green Consumerism. *Journal of Business Strategy*, 13(4), pp. 3–7.

Roberts, J. (1996). Green consumers in the 1990s: Profile and implications for advertising. *Journal of Business Research*, 36(3), pp. 217–231.

Sarigollu, E. (2008). A Cross-Country Exploration of Environmental Attitudes. *Environment and Behavior*, 41(3), pp. 365–386.

Wossen Kassaye, W. (2001). Green dilemma. *Marketing Intelligence and Planning*, 19(6), pp. 444–455.

Author Index

www.ingramcontent.com/pod-product-compliance
Lightning Source LLC
Chambersburg PA
CBHW081645280326
41928CB00069B/2962